An
Economic
Interpretation
of
Linear
Programming

An Economic Interpretation of
LINEAR PROGRAMMING

Quirino Paris

Iowa State University Press / Ames

Quirino Paris is a professor of agricultural economics at the University of California, Davis.

Manufactured in the United States of America on acid-free paper

First edition, 1991

Library of Congress Cataloging-in-Publication Data

Paris, Quirino.
    An economic interpretation of linear programming / Quirino Paris.
  -- 1st ed.
      p.   cm.
    Includes bibliographical references.
    ISBN 0-8138-0469-8 (alk. paper)
    1. Programming (Mathematics)   2. Linear programming.   I. Title.
HB143.P37   1991
330′.01′51972 -- dc20   90-31285
                                                CIP

To the fond memory of

Gerald W. Dean
Lucio De Angelis
and
Vincenzo Cosentino

Contents

Preface

Linear programming has brought to economists the meaning of duality. To be sure, economists have dealt for a long time with production and cost, quantities and prices, but, prior to 1950, the dual relations between technology and profits were known only to a handful of them. Duality, as it turns out, possesses a meaning that transcends linear programming and its economic interpretation.

While searching for the origin of the word, I stumbled upon a book by Boole, who in 1859 wrote: "There exists in partial differential equations a remarkable duality, in virtue of which each equation stands connected with some other equation of the same order by relations of a perfectly reciprocal character." It is not certain whether Boole was the first to introduce the word "duality." The notion, if not the word, however, dates back to 1750 when Euler and Legendre were inventing methods for solving differential equations. Mathematicians, therefore, have been acquainted with duality for at least two centuries. As Boole's words anticipate, the striking feature of duality is made apparent by symmetrical relations that bind the various components of the problem.

Symmetry has always fascinated artists and philosophers. Why should it fascinate mathematicians and scientists? Above all, why should it fascinate economists? The answer is one and the same for everybody: symmetry greatly reduces the set of theoretical specifications that may fit the description of reality. With a smaller number of such specifications, the search and the verification of the most plausible one seem closer at hand. And given the complexity of the reality they study, economists ought to accept symmetry as the principal guideline for their theoretical specifications, on a par with other scientists.

Duality is a gateway to symmetry. We will attempt to illustrate this general statement throughout this book devoted to linear programming, one of the simplest forms of symmetry.

This book began as a set of lecture notes developed over several years for an undergraduate class on linear programming at the University of California at Davis, attended mainly by students of economics and agricultural economics. It requires only a minimal knowledge of linear algebra and calculus. It emphasizes a geometric interpretation of the subject and makes extensive use of diagrammatic illustrations. To fully understand the geometry of the primal and dual spaces, the two most recurring words and notions, the concept of a vector is necessary. All the definitions and algebraic operations required for a serious comprehension of the material are introduced in the main body of the text, avoiding the use of appendices. Duality, symmetry, and economic interpretation are the main focus of the book. Without the direct appeal to duality and symmetry, for example, it would have been exceedingly more difficult to present the dual simplex algorithm of chapter 8. Students' reaction to this chapter has been consistently characterized by disbelief followed by surprised enthusiasm. Personally, I consider it the high point of my linear programming course and book.

The detailed treatment of the simplex algorithms, therefore, is presented to students not merely as a set of recipes for obtaining numerical solutions but, rather, for two more important reasons: because they provide the discipline to study and understand the symmetric relations embodied in linear programming and because every single step and component of the algorithms can be given a clear economic meaning, a feature not easily associated with other procedures. Parametric programming is given a prominent role as a method for deriving economic relations of demand and supply. Unusual attention was paid to the neglected topic of multiple optimal and near optimal solutions. This aspect greatly enhances the ability of linear programming to interpret the structure and the solutions of empirical problems. The last five chapters and the diet problem attempt to illustrate the wide applicability of linear programming, which is bounded, above all, by the user's imagination.

An
Economic
Interpretation
of
Linear
Programming

Chapter 1

Linear Programming

1.1 What Is Linear Programming?

We ask the reader to take the following impromptu quiz: linear programming (LP) is

1. A mathematical technique for solving linear optimization problems subject to linear constraints.
2. A money-making tool.
3. A way of thinking.
4. All of the above.

(Pause for a few seconds to determine your answer before continuing.)

After forty years since the "birth" of linear programming (1947), the correct answer is 4. Indeed, during these years, linear programming went through the three stages indicated by 1, 2, and 3. It all began during the 1940s with the need of the military establishment to make efficient and timely decisions in planning, procurement, scheduling, and deployment. When in 1947, G. B. Dantzig, L. Hurwicz, and T. C. Koopmans discovered a computational procedure (the *simplex method*) for solving a general linear programming problem, the transportation, gasoline blending, and manufacturing industries were ready to apply the new mathematical approach to their decision-making processes. The success, as measured in profit terms, was astonishing according to word-of-mouth reports. Gradually, with the advent of affordable computing facilities, the same success spread from large corporations to firms of any size. Economists were instrumental in this expansion because they were able to explain each step of the linear programming problem and of the simplex algorithm by using a clear, economic terminology.

The principal message of linear programming is that problems come in pairs: the original, almost obvious problem, called the *primal,* and an associated, more subtle problem, called the *dual.* By insisting on the duality aspect of any problem that can be formulated as a linear program, economists propelled linear programming into a way of thinking. There is a special intellectual satisfaction in being able (after a few trials) to organize commonsense information into the structure of a linear programming specification and to anticipate its dual ramifications.

The discovery of linear programming and of the simplex method has forever changed the outlook about the formulation of large-scale problems and about their solution. There seems to be no real limit to the size of problems that can be solved in reasonable time and with affordable cost. Furthermore, a recently discovered procedure, Karmarkar's algorithm, promises to be even more efficient than the simplex method, at least for certain classes of problems, including very large ones.

Hence, there is no lack of motivation for studying linear programming. It represents "practical" mathematics at its best. But, although it is a mathematical scheme obviously useful to theoreticians as well as entrepreneurs, I prefer to emphasize its beauty as expressed through the most elegant symmetry. In point of fact, I strongly hold an unusual viewpoint about it: linear programming is so useful because it is so beautiful.

1.2 Why Another Linear Programming Book?

The nine-campus library system of the University of California lists 490 books under the heading "linear programming." What justifies another book on this subject? Two words: duality and symmetry. In many existing textbooks duality is not given a proper treatment. For economists, on the contrary, duality is the principal source of information about economic relations. The symmetry aspect of linear programming was clearly expounded in early textbooks. I propose, therefore, to devote this entire book to duality, symmetry, and economic interpretation of linear programming and the simplex method. No other algorithm, so far, can be given a more detailed and explicit economic interpretation in all its steps. Furthermore, only by appealing to symmetry considerations is it possible to present the *dual simplex algorithm* and its intricacies in the simple form of chapter 8. Symmetry imposes restrictions that facilitate the explanation and the understanding of difficult problems. This is the deep reason why symmetry, in many different forms, is used in science for interpreting nature.

By using symmetry in economics, we wish to bring the analysis of economic behavior as close as possible to a scientific interpretation.

1.3 Problems for Linear Programming

To provide an immediate illustration of the relevance of linear programming, we discuss two classic, real-life problems that were formulated before any solution methods were available. Indeed, it is safe to assert that the specification of the diet and the transportation problems gave impetus to the search and discovery of a general solution procedure.

The diet problem. In 1945, two years before the appearance of the simplex method, the Nobel-Prize winner George Stigler published a paper in the *Journal of Farm Economics* entitled "The Cost of Subsistence." He wanted to find the minimum-cost diet for a moderately active man. During the WWII period, shortages of various foods appeared throughout the country and concerns were voiced about the adequacy of the diet available to the American people. The debate was fierce, and doubts were raised about the meaning and the feasibility of an adequate diet. Against this background, Stigler formulated one of the first linear programming problems demonstrating the feasibility and the economic meaning of the notion of an adequate diet. To be acceptable, the minimum-cost diet had to satisfy nutritional requirements as specified by the National Research Council and given in table 1.1.

Table 1.1. Daily Allowance of Nutrients for a Moderately Active Man
(weighing 154 pounds)

Nutrient	Allowance
Calories	3,000 calories
Protein	70.0 grams
Calcium	0.8 grams
Iron	12.0 milligrams
Vitamin A	5,000 int. units
Thiamine (B_1)	1.8 milligrams
Riboflavin (B_2 or G)	2.7 milligrams
Niacin (nicotinic acid)	18.0 milligrams
Ascorbic acid (C)	75.0 milligrams

Source: National Research Council, Recommended Dietary Allowances, Circular
Series N. 115, Jan. 1943.

At this point, Stigler compiled the price and the nutritive value of 77 foods as measured by the level of nutrients specified in table 1.1. For that time, it was an overwhelming amount of information, and Stigler proposed to hop toward the solution of the problem by first reducing the list of foods to 15 items, as reported in table 1.2. He then wrote: "Thereafter the procedure is experimental because there does not appear to be any direct method of finding the minimum of a linear function subject to linear conditions." Two years later, a most amazing method for numerical analysis, called the *simplex algorithm,* was available!

The combination of foods eventually selected by Stigler (which could not possibly constitute a true minimum-cost diet) is given in table 1.3.

The real-life problem of choosing an adequate and minimum-cost diet was solved by Stigler using the daily nutrient allowances reported in table 1.1 and then multiplying the results by 365 to obtain annual levels of foods. This procedure is understandable given the computing technology of the time but, of course, it represents a lot of cabbage and beans and, furthermore, fresh spinach was not available all year around. In a later chapter, Stigler's diet problem will be reformulated and solved with the computing facilities of the 1980s.

Linear programming, as Stigler defined it mathematically, is the problem of finding the minimum (or the maximum) of a linear function subject to linear constraints. A miniature diet problem will illustrate this mathematical structure. Suppose the diet problem is simplified to find the minimum cost of a daily diet that must satisfy the nutrient allowance of 3,000 calories and 70 grams of protein.

Table 1.2. Nutritive Values of Common Foods per Dollar of Expenditure, August 1939

Commodity	Unit	Price (cents)	Edible Weight per $1 (grams)	Calories (1,000)	Protein (grams)	Calcium (grams)	Iron (mg)	Vit.A 1,000 (I.U.)	Thiamine (mg)	Riboflavin (mg)	Niacin (mg)	Ascorbic Acid (mg)
1. Wheat flour*	10 lb	36.0	12,600	44.7	1,411	2.0	365	...	55.4	33.3	441	...
2. Evaporated milk*	14.5 oz	6.7	6,035	8.4	422	15.1	9	26.0	3.0	23.5	11	60
3. Oleomargarine	1 lb	16.1	2,817	20.6	17	0.6	6	55.8	0.2	...	...	...
4. Cheese (cheddar)*	1 lb	24.2	1,874	7.4	448	16.4	19	28.1	0.8	10.3	4	...
5. Liver (beef)*	1 lb	26.8	1,692	2.2	333	0.2	139	169.2	6.4	50.8	316	525
6. Green beans	1 lb	7.1	5,750	2.4	138	3.7	80	69.0	4.3	5.8	37	862
7. Cabbage*	1 lb	3.7	8,949	2.6	125	4.0	36	7.2	9.0	4.5	26	5,369
8. Onions	1 lb	3.6	11,844	5.8	166	3.8	59	16.6	4.7	5.9	21	1,184
9. Potatoes	15 lb	34.0	16,810	14.3	336	1.8	118	6.7	29.4	7.1	198	2,522
10. Spinach*	1 lb	8.1	4,592	1.1	106	...	138	918.4	5.7	13.8	33	2,755
11. Sweet potatoes*	1 lb	5.1	7,649	9.6	138	2.7	54	290.7	8.4	5.4	83	1,912
12. Peaches, dried	1 lb	15.7	2,889	8.5	87	1.7	173	86.8	1.2	4.3	55	57
13. Prunes, dried	1 lb	9.0	4,284	12.8	99	2.5	154	85.7	3.9	4.3	65	257
14. Lima beans, dried*	1 lb	8.9	5,197	17.4	1,055	3.7	459	5.1	26.9	38.2	93	...
15. Navy beans, dried*	1 lb	5.9	7,688	26.9	1,691	11.4	792	...	38.4	24.6	217	...

Source: Stigler, G. "The Cost of Subsistence" *Journal of Farm Economics*, 1945, pages 306-7. The asterisk indicates the nine foods eventually selected by Stigler for computing his solution.

Table 1.3. Minimum Cost Annual Diet, August 1939

Commodity	Quantity	Cost
Wheat flour	370 lb	$ 13.33
Evaporated milk	57 cans	3.84
Cabbage	111 lb	4.11
Spinach	23 lb	1.85
Dried navy beans	285 lb	16.80
Total cost		$ 39.93

Source: G. Stigler, "The Cost of Subsistence," *Journal of Farm Economics*, 1945, page 311.

Furthermore, only three types of food are available: wheat flour, potatoes, and lima beans. Let x_1 be the unknown level of wheat flour measured in units of 10 pounds, x_2 be the level of potatoes in units of 1 pound, and x_3 be the level of lima beans in a 1-pound unit. Then, using the information of table 1.2, the minimum-cost diet problem is specified as that of finding nonnegative values of x_1, x_2, and x_3 such that they

(P)   minimize $Z = 36.0x_1 + 2.3x_2 + 8.9x_3$

subject to

$$16.1x_1 + 0.33x_2 + 1.55x_3 \geq 3 \leftarrow \text{calories constraint}$$
$$508.0x_1 + 7.63x_2 + 93.89x_3 \geq 70 \leftarrow \text{protein constraint}$$

$\uparrow$ wheat flour activity $\uparrow$ potatoes activity $\uparrow$ lima beans activity

The corresponding nutritive values of foods per dollar of expenditure as presented in table 1.2 were converted to nutrient values in order to conform with the units of the daily requirements. Calories are measured in units of 1,000.

The first line of problem (P) is the *objective function*, which is linear in the variables x_1, x_2, and x_3. The next two lines are *linear constraints* and specify the main structure of the problem. Since we are not able to interpret negative quantities, we desire to find nonnegative values of x_1, x_2, and x_3, the levels of foods. Problem (P) has very small dimensions: only two nutrient requirements (constraints) and three foods (activities). The original Stigler's problem is a blow-up of problem (P), with 9 nutrient requirements and 77 foods.

The transportation problem. This problem consists of finding the least-cost network of routes for transporting a given commodity from warehouses (plants, origins) to markets (cities, destinations). It has at least two fathers: L. V. Kantorovich and Frank L. Hitchcock. In 1939, the Russian Kantorovich formulated the problem in a rather general form and suggested several applications. In 1941, the American Hitchcock presented his specification. Given those troubled years, the work of Kantorovich reached the West after great delay.

The transportation problem is deceptively simple and serves very well as an illustration of the relevance of linear programming. The idea is to find the minimum cost of transporting, say, potatoes from warehouses to markets. The natural requirements are that each market must receive at least the desired and preordered amount of potatoes and that each warehouse cannot ship more potatoes than are stockpiled; finally, it does not matter which warehouse serves which market as long as the total cost of shipping the potatoes to all markets is the lowest. The cost of shipping a unit of potatoes from each warehouse to each market is fixed. The transportation problem will also be analyzed in greater detail in chapter 4.

1.4   A Profit-Maximizing Problem

The farmer's problem. In this book we will focus on a stylized problem of the competitive farm/firm. Without exaggeration, linear programming is the logical and operational framework that best organizes the amount of complex information for farm/firm decisions. To maximize profits, a farmer must make decisions with respect to the selection of

1. Crop and animal enterprises.
2. Associated operations (tasks) and their scheduling.
3. Capital, machinery, and equipment service flows.
4. Capital, machinery, and equipment stock.
5. Other inputs.

Even for a modest farm, the amount of information to collect and process for articulating a production plan is far from trivial. Here is where linear programming can be of valuable assistance. To obtain good results, it must be used correctly, with imagination and inventiveness, and must be fed accurate information. It must be kept in mind, therefore, that the quality of results obtained from a linear programming solution cannot be better than the original information fed into the model. Hence, if the solution seems implausible, the linear programming approach should not be blamed, only its user.

1.5   A First Encounter with Duality

Most problems possess a twofold representation that is customarily defined as *duality*. Duality is nothing else than the specification of *one* problem from *two* different points of view. Once a problem is specified, we refer to it as the *primal* problem. This primal formulation is, in general, obvious, since it represents the problem as we wish to describe it and are capable of doing so.

A second, associated problem usually exists and is called the *dual* (from the Latin adjective "duo," which literally means "two" but which here can better be translated as "second") specification. This dual specification is not obvious and therefore conveys an unexpected amount of information. Economists are especially interested in the dual problem (more, if possible, than in the primal problem) because it contains mostly economic information.

We will begin the study of duality by specifying the classical problem of the competitive firm. This way of approaching duality is rigorous and sufficiently general.

The farm decision problem in a perfectly competitive environment can be stated as the following primal problem:

(P)   maximize total revenue = TR

   subject to the production technology, or (more explicitly)
   total input use $\leq$ total available inputs

Total revenue is defined as the sum of all product quantities multiplied by their respective (fixed) prices. Total input use is defined as the input requirement per unit of output multiplied by the output level and summed over all products. The constraints of problem (P) represent the production function of economic theory in an unfamiliar specification. In order to make the necessary connection, the production function is usually stated in terms of outputs as a function of inputs, that is,

$$\text{outputs} = f(\text{inputs})$$

In linear programming, the same production function is stated in a reciprocal form as

$$h(\text{outputs}) = \text{inputs}$$

Although the above distinction between inputs and outputs may be convenient for some simplified problem, it is wise to keep in mind that a commodity can be both an input *and* an output within the same firm.

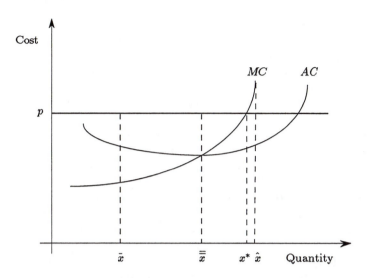

Figure 1.1. The average and marginal cost functions with a nonlinear technology.

Hence, in a really general way, we ought to be speaking of producing commodities by means of commodities. The technology, in other words, relates commodities to commodities, with the consequence that the same commodity may appear on both sides of the equations (inequalities) as an output *and* as an input.

Associated with the primal (P) there exists a very important problem, called the dual, which is specified as

(D)   minimize total input costs $= TC$

subject to equilibrium conditions, or (more explicitly)
$MC =$ marginal activity cost $\geq$ marginal revenue $= MR$

In familiar economic theory courses, the firm's equilibrium condition is presented as an equality, that is,

marginal cost $=$ marginal revenue

Problem (D), however, exhibits an important generalization of this equilibrium condition. But why should the general relation between MC (marginal cost) and MR (marginal revenue) be as $MC \geq MR$ rather than $MC \leq MR$? An examination of figure 1.1 will clarify the dilemma.

Figure 1.1 represents the diagram relating average and marginal cost curves to the competitive output price $(p = MR)$ for an underlying nonlinear technology (production function). We wish to establish the condition under which a firm is in equilibrium, or, in other words, we wish to make the determination of the optimal quantity of output to produce. The idea here is that a firm is in equilibrium when the optimal product mix has been ascertained. That situation represents an equilibrium for the profit maximizing firm because it has no interest in deviating from that product mix.

Consider the quantities $\bar{x}$ and $\bar{\bar{x}}$ in figure 1.1. In both cases, $MC < p$ and it is clear that an extra unit of output will generate extra profit for the firm. Therefore, neither quantity can represent an equilibrium output. This conclusion can be extended to all output quantities for which $MC < p$. Having excluded the relationship $MC < p$, the general condition for the firm's equilibrium is $MC \geq p$. To fully comprehend this equilibrium condition, it is convenient to specialize figure 1.1 to represent the same relations between average and marginal cost curves for linear technologies, since linear programming presupposes such production functions.

It is a fact that, for a linear production function, the average and marginal cost curves coincide for any output quantity. This condition can occur if and only if average and marginal cost "curves" are straight horizontal lines (see section 1.8 for a mathematical clarification). Figure 1.2, then, represents possible relations (associated with four different linear technologies) between marginal cost, MC (and, implicitly, average cost, AC), and marginal revenue, MR, which is equal to the market price, p, in a competitive environment. There are three cases: either $MC < MR$, or $MC = MR$, or $MC > MR$.

We concluded before that, when $MC < MR$, it is convenient to further expand output production. Figure 1.2 shows that, under this condition, a positive profit accrues and, therefore, it cannot represent an equilibrium point. When $MC > MR$, a loss occurs for each unit of output. It is convenient, therefore, to produce zero quantity of output. The *equilibrium* quantity in this case is $x^* = 0$. Another equilibrium quantity, corresponding to a positive output, is when $MC = MR$, the familiar condition.

Although figure 1.2 would seem to indicate that, in this case, output should be expanded indefinitely, it is actually blocked at x^* by the constraints representing the limiting resources.

To summarize, only when $MC \geq MR$ do we have an equilibrium for the firm. If $MC = MR$, it is convenient to produce the associated output quantity. When $MC > MR$, the guideline is that the optimal output quantity is zero, $x^* = 0$. To reinforce this point, suppose a farmer for the East Coast comes to Yolo County, California, to set up a new farm operation. He knows that a large number of crops can be cultivated in California. Hence, in his first season and with current output prices, he attempts to produce wheat,

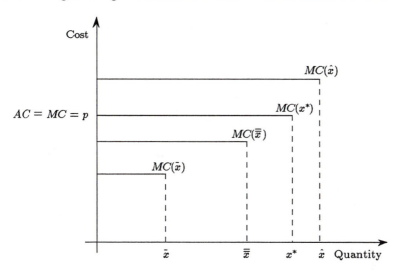

Figure 1.2. The average and marginal cost functions with a linear technology.

corn, alfalfa, rice, lettuce, and peanuts. He soon discovers (at the end of the season) that his marginal costs for lettuce and peanuts are greater than the corresponding market prices. Thus, he decides to eliminate those two crops from next year's production plan.

1.6 Profit Maximization in LP: The Primal Problem

The same two problems, (P) and (D), are now stated in a linear programming framework (with only two outputs and two inputs). The primal problem (P) is respecified as:

(P′) maximize $Z = p_1 x_1 + p_2 x_2 \equiv$ total revenue

$$
\begin{array}{rcccl}
\text{subject to} \quad 2x_1 & + & 6x_2 & \leq & 12 \leftarrow \text{land constraint} \\
4x_1 & + & x_2 & \leq & 8 \leftarrow \text{labor constraint} \\
x_1 & & & \geq & 0 \\
& & x_2 & \geq & 0
\end{array}
$$

$\uparrow$ wheat production activity $\uparrow$ corn production activity where x_1 and x_2 are the (unknown) output levels of wheat and corn; p_1 and p_2 are their corresponding competitive market prices, while $\{2, 6, 4, 1\}$ constitute the technical input requirement coefficients or the production technology (production function).

The coefficient 2, for example, represents the units of land necessary for producing one unit of wheat. The land coefficients are usually chosen to be of unit value, implying that all the other coefficients for the same activity, say wheat, are measured with reference to a unit of land, say an acre. In this first example, we deliberately choose a nonunit coefficient for land to alert the reader to the important problem of measurement units, which will be further discussed in chapter 3. In Europe, for example, the hectare is the conventional unit for measuring land and it transforms into about 2.2 acres via a divisor coefficient of 0.45. Similarly, the coefficient 4 in the constraints of problem (P′) indicates that for producing one unit of wheat it is necessary to use 4 units of labor.

To perceptive readers it may seem that to go from a nonlinear production function (as in the usual economic theory courses) to a linear technology (as in linear programming) entails a considerable loss of generality. There is no doubt that their perception is correct only to be substantially mitigated by the realization that with linear programming we can immediately deal with multiple outputs and multiple limiting ("fixed") inputs without the complexities of a nonlinear mathematical specification. Furthermore, it is often possible to approximate very well a nonlinear function by a sequence of linear segments. Hence, linear programming is not as restrictive as it may appear at first glance.

1.7 Free Goods

Before proceeding with the derivation of the dual problem (D′) corresponding to (P′), it is convenient to introduce the notion of a free good. It is well known that, in competitive markets, commodities command positive prices only when they are scarce relative to their demand. Linear programming is based upon the notion of marginal pricing in a competitive environment. According to this scheme, all units of a resource good should be priced in proportion to its corresponding marginal product (MP). If, for a given resource (input), $MP = 0$, the (imputed, economic) price of that resource is also zero. In turn, $MP = 0$ when some units of the resource in question are left unused. Hence, when the supply of an input is strictly greater than the use (demand) of that input, its imputed (economic, as distinct from accounting) price is equal to zero. This pricing scheme is based upon the notion of the availability of alternative uses for the given resource. When the entire supply of a resource cannot be employed in a particular line of production and, furthermore, alternative opportunities for the employment of the surplus quantity do no exist, the last (marginal) unit of that resource is associated with a zero marginal product and, thus, its price is zero. This means also that the units of that resource used up in production will be remunerated at the same zero price, a conclusion that often causes a great deal of consternation to the beginning LP student.

1.8 Cost Minimization in LP: The Dual Problem

We now want to set up the dual problem (D′) corresponding to the primal problem (P′). To do so, we will refer to the general specification of the firm problem as stated in problem (D). That problem requires the computation of the marginal cost and marginal revenue for each output. Notice that in (P′) the quantities to the left-hand side of the constraint inequalities are interpreted as

$$(2x_1 + 6x_2) \equiv \text{total use (demand) of input land}$$
$$(4x_1 + x_2) \equiv \text{total use (demand) of input labor}$$

If we let y_1 and y_2 be the imputed prices of land and labor, respectively, the total cost (TC) of those inputs must be the sum of the input quantities times their prices, that is,

$$
\begin{aligned}
TC &= \text{(total input uses)} &\times& \text{ (imputed prices)} \\
&= (2x_1 + 6x_2)y_1 &+& (4x_1 + x_2)y_2 = 12y_1 + 8y_2 \\
&= (2y_1 + 4y_2)x_1 &+& (6y_1 + y_2)x_2 = 12y_1 + 8y_2 \\
&= TC \text{ of wheat} &+& TC \text{ of corn } = 12y_1 + 8y_2 \\
&= \text{(input supplies)} &\times& \text{ (imputed prices)}
\end{aligned}
$$

We began by stating the definition of total cost in an obvious manner (input quantities multiplied by their prices) both in English and in mathematical terms. Then, on the third and fourth lines, a simple rearrangement of the variables produced another equivalent specification of total cost in terms of output quantities multiplied by their average (and marginal) costs. The equivalence between [(total input uses) × (imputed prices)] and [(input supplies) × (imputed prices)] expressed in the total cost equation is based upon the notion of a free good. In fact, if the input use (demand) happens to be strictly less than input supply, it is a free good and the corresponding imputed price must be zero. In general, therefore, the price of an input is strictly positive only when the input supply is equal to its demand (use). For this reason, the total cost can be expressed in the two equivalent forms indicated above.

Let us recall that marginal cost is the first derivative of total cost with respect to the output variable. Since in this example there are two outputs, wheat and corn, there are also two marginal costs. Hence,

$$\text{for output 1:} \quad \frac{\partial TC}{\partial x_1} = 2y_1 + 4y_2 \equiv MC_1$$

$$\text{for output 2:} \quad \frac{\partial TC}{\partial x_2} = 6y_1 + y_2 \equiv MC_2$$

Notice that MC_1 does not depend on x_1 and, therefore, it is a constant (see figure 1.2) depending only on the values of y_1 and y_2 which do not appear in figure 1.2. Also notice that, from the expression of total cost, we can rewrite TC as

$$TC = TC \text{ of } x_1 + TC \text{ of } x_2$$

Hence, the average cost of x_1, (for $x_1 > 0$), is equal to the marginal cost of x_1, as asserted:

$$AC_1 = \frac{TC \text{ of } x_1}{x_1} = 2y_1 + 4y_2 = MC_1$$

Analogously, marginal revenue is the first derivative of total revenue with respect to output quantities. Therefore,

$$\text{for output 1:} \quad \frac{\partial TR}{\partial x_1} = p_1 \equiv MR_1$$

$$\text{for output 2:} \quad \frac{\partial TR}{\partial x_2} = p_2 \equiv MR_2$$

We are now ready to assemble the dual problem (D′) corresponding to the primal problem (P′) following the specification given in (D):

(D′)   minimize $TC = 12y_1 + 8y_2$

$$
\begin{array}{lrcccll}
\text{subject to} & 2y_1 & + & 4y_2 & \geq & p_1 & \leftarrow \text{wheat price constraint} \\
 & 6y_1 & + & y_2 & \geq & p_2 & \leftarrow \text{corn price constraint} \\
 & y_1 & & & \geq & 0 & \\
 & & & y_2 & \geq & 0 & \\
\end{array}
$$

$$
\begin{array}{cc}
\uparrow & \uparrow \\
\text{land} & \text{labor} \\
\text{cost} & \text{cost} \\
\text{activity} & \text{activity}
\end{array}
$$

An interesting transposition of coefficients representing the technical information has occurred between problems (P′) and (D′). The technological information of the primal problem (P'), expressed by the matrix

$$\begin{bmatrix} 2 & 6 \\ 4 & 1 \end{bmatrix}$$

appears transposed in the dual problem (D'), where the above matrix appears as

$$\begin{bmatrix} 2 & 4 \\ 6 & 1 \end{bmatrix}$$

This general property of linear programming is a consequence of duality. It means that the inequalities that represent the linear constraints can be read in two ways: as rows and as columns. In fact, we have anticipated this property of linear programming in the above specification of the primal and the dual problems.

The prices p_1 and p_2 appear as coefficients of the objective function in the primal problem (P') and as constraint coefficients in the dual problem (D'). The land and labor availabilities (12 and 8) are constraint coefficients in the primal problem and objective function coefficients in the dual. Finally, the direction of the constraint inequalities is reversed in the two problems. The remarkable aspect of this dual specification is that we have used only economic information to correctly set up a pair of related linear programming problems.

To minimize confusion in setting up dual problems, it is important to notice one crucial point: the primal inequalities of the structural constraints (such as land and labor, in the above example) were all organized in one direction ($\leq$); as a consequence, the dual inequalities were also all organized in one (the opposite) direction ($\geq$). This mnemonic rule has its justification grounded in economic theory, as illustrated above. Often, however, primal problems can be specified with inequalities running in either direction ($\leq$ and/or $\geq$). In this case, to avoid setting up dual problems that are either misspecified or difficult to interpret, it is very convenient to rewrite the primal constraints so that all of them exhibit inequalities running in the same direction. By strictly following this rule, the dual problem will be correctly set up and will be easily interpreted. In this vein of thought, it is also convenient to associate an objective function to be maximized with less-than-or-equal ($\leq$) constraints and a minimization with greater-than-or-equal ($\geq$) constraints.

One further look at the specification of the primal and dual problems (P') and (D'): the complex of transpositions indicated above, the structure of the constraints, and the independent appearance of the variables constitute a *symmetry* that plays a fundamental role in linear programming. Symmetry and duality, therefore, are two related properties that

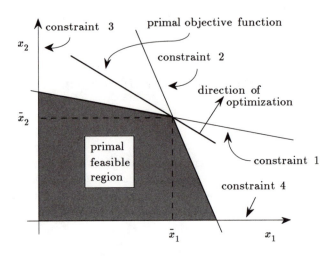

Figure 1.3. The primal optimal solution.

must gradually be understood to fully appreciate the potential of mathematical programming.

1.9   Graphical Solution of LP Problems

The specification of problems (P′) and (D′) was purposely made simple (4 constraints and 2 variables) to be able to use graphical methods in finding the corresponding optimal solutions. To solve the primal problem (P′), it is sufficient to graph the primal constraints in the output space (the space generated by the output levels, x_1 and x_2) indicating the correct direction of the inequalities with a perpendicular arrow. Later on we will see that this arrow can be interpreted as the vector of coefficients defining the corresponding constraint. This procedure identifies the feasible region of the primal problem. Next, we graph the free-floating primal objective function and slide it parallel to itself in the direction of optimization until it touches the feasible region at a corner point. This is the optimal extreme point corresponding with the solution of the problem. It remains to read off the optimal output levels by dropping a perpendicular (dashed) line from the optimal extreme point (corner point) to the coordinate axes. This procedure is illustrated in figure 1.3.

The output quantities $\bar{x}_1$ and $\bar{x}_2$ are part of the primal optimal solution. The rest of the primal solution is given by the values of the slack variables to be defined in section 1.11. The optimal value of the primal

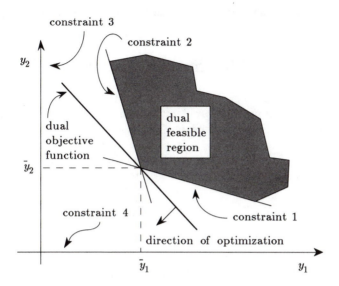

Figure 1.4. The dual optimal solution.

objective function is easily computed by evaluating the primal solution as in the following quantity

$$p_1 \bar{x}_1 + p_2 \bar{x}_2 = \overline{TR}$$

A similar procedure is applied to the dual problem, as illustrated in figure 1.4. It is important to note that the dual inequality constraints must be graphed in the correct direction. Also the dual objective function is slid in a direction opposite to that of the primal problem.

The values $\bar{y}_1$ and $\bar{y}_2$ are the optimal imputed prices of the land and labor inputs, respectively. Together with the values of the dual slack variables they constitute the dual optimal solution. The optimal value of the dual objective function is easily computed as

$$12\bar{y}_1 + 8\bar{y}_2 = \overline{TC}$$

Notice that the dual feasible region, contrary to the primal feasible region, is open in the northeast direction. This fact, however, does not cause any difficulty in this example because the direction of optimization is pointed toward the southwest, where the region is bounded. Comparing figures 1.3 and 1.4, we can conclude that although the primal and dual problems are symmetric in the sense explained above, they are not identical.

The two figures, therefore, provide an important insight into the meaning of symmetry.

1.10 Additional Information: The Dual of the Dual

It is important to emphasize that a primal problem could very well be specified as a minimization problem; in such a case, its corresponding dual would be a maximization problem. The diet and the transportation problems stated in section 1.3, are in fact stated as minimizing primal problems. The farmer's problem could also have been initially stated as a minimization. For example, we could have chosen to specify the primal problem (P') as

(P*)   minimize $TC = 12y_1 + 8y_2$

subject to

$$
\begin{array}{rcrcll}
2y_1 & + & 4y_2 & \geq & p_1 & \leftarrow \text{ wheat price constraint} \\
6y_1 & + & y_2 & \geq & p_2 & \leftarrow \text{ corn price constraint} \\
y_1 & & & \geq & 0 & \\
& & y_2 & \geq & 0 &
\end{array}
$$

$$
\begin{array}{cc}
\uparrow & \uparrow \\
\text{land} & \text{labor} \\
\text{cost} & \text{cost} \\
\text{activity} & \text{activity}
\end{array}
$$

To set up the dual problem (D*) corresponding to a cost minimization, we must obtain the total revenue (TR) function, the input use relations (technology), and the supply of inputs using only the information of the primal problem (P*). Recall that the total revenue is the integral under the marginal revenue function, which in turn is stated as

$$\frac{\partial TR}{\partial x_1} = MR_1 = p_1$$

Hence, the total revenue due to output x_1 is

$$TR_1 = \int_0^{x_1^*} p_1 \, dx_1 = p_1 x_1^*$$

Combining TR_1 and TR_2 (similarly derived), the total revenue is

$$TR = TR_1 + TR_2 = p_1 x_1 + p_2 x_2$$

Notice that the total cost function, TC, is already stated in (P*) as $TC = 12y_1 + 8y_2$, but analogously to the TR function, we can use the marginal cost functions specified in the constraints of (P*) to derive the total cost function. Therefore, the total cost function relating to output 1 is the integral under the marginal cost function corresponding to that output:

$$TC_1 = \int_0^{x_1^*} (2y_1 + 4y_2)dx_1 = 2x_1^*y_1 + 4x_1^*y_2$$

Similarly,

$$TC_2 = \int_0^{x_2^*} (6y_1 + y_2)dx_2 = 6x_2^*y_1 + x_2^*y_2$$

Hence, total cost is

$$
\begin{aligned}
TC &= TC_1 & + \; TC_2 \\
&= (2x_1y_1 + 4x_1y_2) & + \; (6x_2y_1 + x_2y_2) \\
&= (2x_1 + 6x_2)y_1 & + \; (4x_1 + x_2)y_2 &= 12y_1 + 8y_2 \\
&= \text{(input uses)} \times \text{(imputed prices)}
\end{aligned}
$$

As for TR and TC, the asterisk is left out to signify that these relations are valid for any nonnegative level of x_1 and x_2, not only for x_1^* and x_2^*.

The student of economics has been exposed to the process of taking derivatives of the total cost and total revenue functions with respect to output quantities. Those derivatives are called marginal cost and marginal revenue, respectively. One interesting question, however, is whether there is any meaning in the derivative of total cost with respect to input prices. As will be shown below, this is a dual question!

Using the specification of the total cost function just given, let us first compute the derivatives with respect to (imputed) prices and then interpret them:

$$\frac{\partial TC}{\partial y_1} = 2x_1 + 6x_2 \equiv \text{ use (demand) of input 1}$$

$$\frac{\partial TC}{\partial y_2} = 4x_1 + \; x_2 \equiv \text{ use (demand) of input 2}$$

We recognize, therefore, that the derivatives of TC with respect to input prices result in the respective demand functions. (Confront the results with the constraints of the primal problem (P′)). Analogously, if we use the total cost function given in (P*),

$$\frac{\partial TC}{\partial y_1} = 12 \equiv \text{supply (availability) of input 1}$$

$$\frac{\partial TC}{\partial y_2} = 8 \equiv \text{supply (availability) of input 2}$$

we obtain the corresponding input supplies. Remarkably, we have derived the components of the dual problem (D*) in a way entirely analogous to that used for obtaining the dual (D').

We can now proceed to assemble the various components into the dual problem:

(D*) maximize $Z = p_1 x_1 + p_2 x_2 \equiv$ total revenue

subject to

land demand $\rightarrow$	$(2x_1$	$+$	$6x_2)$	$\leq 12 \leftarrow$ land supply
labor demand $\rightarrow$	$(4x_1$	$+$	$x_2)$	$\leq 8 \leftarrow$ labor supply
	x_1			≥ 0
			x_2	≥ 0

$$\uparrow \qquad\qquad \uparrow$$

wheat   corn
production   production
activity   activity

which we recognize to be identical to the primal problem (P'). We conclude that for this class of problems *the dual of the dual is the primal.*

1.11 Slack (Surplus) Variables and Their Meaning

The constraints of the primal problem as written in (P') exhibit inequalities that are interpreted as

input use (demand) $\leq$ input availability (supply)

In this case, it is always possible and easy to transform the inequality into an equation by defining (for each inequality) an additional variable, called the *slack* or *surplus* variable, as follows:

primal slack variable = input availability − input use ≥ 0

Defined in this way, the slack (or surplus) variable is nonnegative. The name "slack" variable indicates that its main role is that of taking up the slack between the right-hand side and the left-hand side of the inequality sign. The name "surplus" variable signifies the difference between the input availability and the input use. As an illustration, the two constraints of problem (P') can be rewritten (without modifying their meaning) as

$$
\begin{aligned}
2x_1 + 6x_2 + x_{s1} \qquad\qquad &= 12 \\
4x_1 + x_2 \qquad\quad + x_{s2} &= 8 \\
x_1 \qquad\qquad\qquad &\geq 0 \\
x_2 \qquad\qquad\quad &\geq 0 \\
x_{s1} \qquad\quad &\geq 0 \\
x_{s2} &\geq 0
\end{aligned}
$$

where x_{s1} and x_{s2} are the slack or surplus variables corresponding to the land and labor constraints, respectively. Surplus variables are, therefore, associated with a no production activity or, equivalently, with an idle amount of resource.

The constraints of the dual problem (D') can also be associated with a slack (or marginal loss) variable, which has the following definition and meaning:

$$
MC = MR + \text{dual slack variable}
$$

or

$$
MC - \text{marginal loss} = MR
$$

where marginal loss is identically equal to the dual slack variable.

Using the constraints of (D'), we can implement this last definition of dual slack as

$$
\begin{aligned}
(2y_1 + 4y_2) - y_{s1} \qquad\qquad &= p_1 \\
(6y_1 + y_2) \qquad\quad - y_{s2} &= p_2 \\
y_1 \qquad\qquad\qquad &\geq 0 \\
y_2 \qquad\qquad\quad &\geq 0 \\
y_{s1} \qquad\quad &\geq 0 \\
y_{s2} &\geq 0
\end{aligned}
$$

where $y_{s1} \equiv$ marginal loss for commodity 1 and $y_{s2} \equiv$ marginal loss for commodity 2. Note that, in the dual constraints, the dual (nonnegative) slack variables are subtracted, rather than added as in the primal constraints. This is a consequence of adhering to the convention of keeping all the variables on one side (the left-hand side) of the equality sign. In the primal problem, the nonnegative slack variables are added to the input demand, which is always less than or equal to input supply. In the dual problem, the nonnegative marginal loss is subtracted from the marginal cost, which is always greater than or equal to marginal revenue.

1.12   A General Linear Programming Formulation

Occasionally, it will be useful to discuss the structure of linear programming problems free from the shackles of Arabic numbers. For this purpose, we define

c_j ≡ unit net revenue of commodity j
b_i ≡ total available quantity of input i
a_{ij} ≡ amount of input i necessary for the production of one unit of commodity j
x_j ≡ level of commodity j produced using the jth technological process
y_i ≡ imputed (shadow) price of input i

The general formulation of a primal LP problem (with a maximization chosen for no essential reason) can be specified as follows (TNR = total net revenue):

$$\text{maximize } TNR = \sum_{j=1}^{n} c_j x_j$$

$$\text{subject to} \quad \sum_{j=1}^{n} a_{ij} x_j \le b_i, \qquad i = 1, \ldots, m$$

$$x_j \ge 0, \qquad j = 1, \ldots, n$$

The corresponding dual problem is

$$\text{minimize } TC = \sum_{i=1}^{m} b_i y_i$$

$$\text{subject to} \quad \sum_{i=1}^{m} a_{ji} y_i \ge c_j, \qquad j = 1, \ldots, n$$

$$y_i \ge 0, \qquad i = 1, \ldots, m$$

The symmetric duality of a general linear programming problem can be clearly exhibited by arranging all the information of the primal and the dual problems in a single array (adapted from Goldman and Tucker):

≥ 0	x_1	x_2	$\ldots$	x_n	$\le$
y_1	a_{11}	a_{12}	$\ldots$	a_{1n}	b_1
y_2	a_{21}	a_{22}	$\ldots$	a_{2n}	b_2
$\ldots$	$\ldots$	$\ldots$	$\ldots$	$\ldots$	$\ldots$
y_m	a_{m1}	a_{m2}	$\ldots$	a_{mn}	b_m
$\ge$	c_1	c_2	$\ldots$	c_n	$\le$

The primal problem can be read by multiplying the variables x_j to the coefficients in the corresponding columns. Similarly, the dual problem can be identified by multiplying the y_i variables across the corresponding rows. The inequality in the northwestern corner represents the nonnegative values of both primal and dual variables. The inequalities in the northeast and southwest locations represent the primal and the dual constraints, respectively. Finally, the inequality in the southeast corner represents the relationship between the primal and the dual objective functions.

1.13   Three Crucial Assumptions

Linear programming relies on the assumptions of *proportionality, additivity,* and *divisibility.*

Proportionality. The assumption asserts that, if it requires 1 unit of land and 3 units of labor to produce 40 bushels of wheat, it will take 2 units of land and 6 units of labor to double the wheat output. The generalization of this example suggests that an activity can be operated as a multiple of the resource requirements specified for a unit level of output.

Additivity. The assumption states that, if the production of unit levels of wheat and corn require land and labor in different amounts, the demand of land is equal to the sum of the different quantities of land demanded for the production of wheat and corn. Similarly, the demand of labor is equal to the sum of the different quantities of labor demanded for the production of wheat and corn. The additivity assumption implies that both the constraints and the objective function of a linear programming problem are (additively) separable in the primal variables.

Divisibility. The assumption asserts that every (primal and dual) variable can take on any real value compatibly with the specified constraints. When dealing with lumpy commodities such as machinery, animals, buildings, etc., this assumption may not be satisfactory. In order to correctly specify lumpy commodities, integer programming is required. Integer programming will not be discussed in this book.

Chapter 2

Primal and Dual LP Problems

2.1 The Structure of Duality

The notion of duality plays a very important role in linear programming. Indeed, it also plays a crucial role in nonlinear programming, economics, engineering, physics, and statistics. It is possible to assert that almost every problem has a primal and a dual specification.

Duality is particularly important for economists because almost all the economic information comes from the dual problem. Furthermore, duality is beautiful, symmetric, and useful. It is useful because it is symmetric and, thus, beautiful. This aspect of the scientific method is worth emphasizing. It is important to look at the symmetry (often, a hidden symmetry) of problems in order to discover their usefulness. The converse is not so important or even true: searching directly for the practicality of problems is not easy. It appears, therefore, that usefulness is a roundabout property of problems that is more easily brought to the surface by symmetric specifications.

The reader should seriously think of duality as one of the most important concepts heard in life. As we hope to show in this book, it is worth spending the necessary time to learn such a beautiful and general principle of nature and mathematics because the dual problem is always stunningly surprising and unexpectedly informative.

In chapter 1 we observed that the dual of the dual problem is the primal problem. This reflexive property of linear programming allows us to speak of the primal and dual problems simply as a pair of dual problems. In spite of the fact that no dual problem would exist without its primal specification, the above terminology emphasizes the importance of duality over the notion of "primality." Indeed, we will never speak of a "primal pair" of linear programming problems.

To fully understand the duality relations in linear programming, it is important to know how to set up dual pairs of problems properly and how to interpret all their components in economic terms. The general economic framework for the competitive firm specified in chapter 1 resulted in the following dual pair of problems:

Primal	**Dual**
max NR = net revenue	min TC = total cost
subject to production technology	subject to $MC \geq MR$

We wish to remark that this framework is very general and, in fact, it can be taken as the definition of duality itself without fear of being too restrictive. The student of economics, therefore, has already encountered duality.

By restricting the general framework to a linear technology, it is possible to restate the dual pair of problems as

<table>
<tr><td colspan="4" align="center">Primal</td><td colspan="2" align="center">Dual</td></tr>
<tr><td colspan="4">max $NR = 30x_1 + 45x_2$</td><td colspan="2">min $TC = 15y_1 + 10y_2 + 0y_3$</td></tr>
<tr><td colspan="4">subject to</td><td colspan="2">subject to</td></tr>
<tr><td>input use</td><td>$\leq$</td><td colspan="2">input supply</td><td colspan="2">imputed input price $\geq$ 0</td></tr>
<tr><td>$4x_1 + 3x_2$</td><td>$\leq$</td><td>15</td><td>$\longleftrightarrow$</td><td>y_1</td><td>$\geq$ 0</td></tr>
<tr><td>$2x_1 + x_2$</td><td>$\leq$</td><td>10</td><td>$\longleftrightarrow$</td><td>y_2</td><td>$\geq$ 0</td></tr>
<tr><td>$-x_1 + 5x_2$</td><td>$\leq$</td><td>0</td><td>$\longleftrightarrow$</td><td>y_3</td><td>$\geq$ 0</td></tr>
<tr><td>x_1</td><td>$\geq$</td><td>0</td><td>$\longleftrightarrow$</td><td>$4y_1 + 2y_2 - y_3$</td><td>$\geq$ 30</td></tr>
<tr><td>x_2</td><td>$\geq$</td><td>0</td><td>$\longleftrightarrow$</td><td>$3y_1 + y_2 + 5y_3$</td><td>$\geq$ 45</td></tr>
<tr><td>activity levels</td><td>$\geq$</td><td>0</td><td>$\longleftrightarrow$</td><td colspan="2">$MC \geq MR$</td></tr>
</table>

Notice the pairing of primal and dual components: to primal (technological) constraints there correspond dual variables (imputed prices); to primal variables (activity levels) there correspond dual constraints (economic equilibria). The converse is also true: to dual variables there correspond primal constraints; to dual constraints there correspond primal variables. This makes sense on economic grounds because to a primal constraint that represents an input availability there corresponds a dual variable that represents the imputed (shadow) price of that input; to an activity level there corresponds a marginal cost–marginal revenue relation that is the main criterion in assessing whether that activity is profitable. In a more general sense: to the primal constraints that represent the structure of technology there corresponds a system of input prices; to the set of primal variables that represents a production schedule there corresponds a set of dual constraints that represents the cost-revenue structure for producing those planned commodities.

In mathematical terms, the duality relations are as follows:

1. To a primal constraint there corresponds a dual variable (never a dual constraint).

2. To a dual variable there corresponds a primal constraint (never a primal variable).

3. To a primal variable there corresponds a dual constraint (never a dual variable).

4. To a dual constraint there corresponds a primal variable (never a primal constraint).

Furthermore, notice that to inequality constraints in the primal problem there correspond nonnegative dual variables, and to nonnegative primal variables there correspond inequality constraints in the dual problem.

These dual relations are essential for understanding the structure and the working of linear programming as well as the corresponding economic interpretation. The insistence on the proper pairing of dual items is justified by the possibility of an erroneous association of primal variables with dual variables and of primal constraints with dual constraints. These combinations cannot occur for several reasons. The most obvious one is that, in general, a linear programming problem is specified with a different number of primal and dual constraints, as in the above example. Similarly, there are unequal numbers of primal and dual variables. It is, therefore, unthinkable to match primal and dual constraints and primal and dual variables. A second reason is based on the economic interpretation of a LP problem. Using the framework of the competitive firm, the primal variables might be interpreted as output levels and the dual variables as input prices. There is no admissible rationale for matching these two kinds of information.

2.2   Dual Pairs of LP Problems with Equality Constraints

The linear programming examples formulated in previous sections presented only inequality constraints. Empirical problems, however, come in all shapes and forms, and it is not unusual to encounter linear programming problems with inequality constraints as well as equations. A legitimate question, then, concerns the proper relation between constraints in the form of equations and the associated dual variables. In other words,

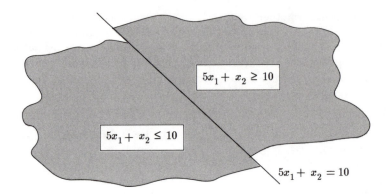

Figure 2.1. A line as the intersection of two closed half spaces.

while we already know that to primal constraints in the form of inequalities there correspond nonnegative dual variables, what can be said about the dual variable associated with a constraint in the form of an equation? In this section, we will utilize the information already at hand to answer this important question.

We begin by noticing that, in general, an equation can be equivalently specified as two inequality constraints. For example,

$$(2.1) \qquad 5x_1 + x_2 = 10 \longleftrightarrow \begin{cases} \text{(a) } 5x_1 + x_2 \leq 10 \\ \text{(b) } 5x_1 + x_2 \geq 10 \end{cases}$$

Geometrically, a constraint such as equation (2.1) represents a line. On the contrary, a constraint such as the inequality (2.1.a) represents a closed half space: in this case, one half of a plane, as illustrated in figure 2.1. Similarly, inequality (2.1.b) represents the other half space. We conclude that the intersection of two complementary inequality constraints is equal to an equation and, therefore, a line can be equivalently represented as either in (2.1) or in (2.1.a) and (2.1.b).

Consider now the following primal LP problem:

Primal maximize $NR = 30x_1 + 45x_2$

$$\begin{aligned}
\text{subject to} \qquad 4x_1 + 3x_2 &= 15 \\
2x_1 + x_2 &\leq 10 \\
-x_1 + 5x_2 &\leq 0 \\
x_1 &\geq 0 \\
x_2 &\geq 0
\end{aligned}$$

The peculiarity of this example is that the first constraint is an equation. From chapter 1 we know that to inequality constraints there correspond nonnegative dual variables. What can be said about the dual variable corresponding to a constraint in the form of an equation, as in the above example? To answer this question we will use only the knowledge of how to set up linear programming problems already developed in chapter 1. Our problem, therefore, is to write down the correct dual corresponding to the above primal problem while following the specifications given in chapter 1, which are reproduced here for convenience: the coefficients of the primal objective function become the dual constraints; the primal constraints become the coefficients of the dual objective function; each column of coefficients in the primal constraints becomes a row of coefficients in the dual constraints; inequalities are reversed and variables are nonnegative. Finally, to a maximization problem subject to less-than-or-equal constraints there corresponds a minimization problem subject to greater-than-or-equal constraints. This list of rules is only a mnemonic way to look at a pair of dual linear programming problems. In chapter 1 we have discussed the reasons for such a symmetry.

To answer the question formulated at the start of this section, let us replace the equality constraint with its corresponding two inequalities.

Primal maximize $NR = 30x_1 + 45x_2$

$$
\begin{array}{rrclcl}
 & & & & & \text{dual} \\
 & & & & & \text{variables} \\
\text{subject to} \quad 4x_1 & + & 3x_2 & \leq & 15 & y_1' \\
-4x_1 & - & 3x_2 & \leq & -15 & y_1'' \\
2x_1 & + & x_2 & \leq & 10 & y_2 \\
-x_1 & + & 5x_2 & \leq & 0 & y_3 \\
x_1 & & & \geq & 0 & \\
 & & x_2 & \geq & 0 &
\end{array}
$$

The problem now conforms to our knowledge of the relationship between inequality constraints and nonnegative dual variables. The name chosen for the dual variables of the first two inequalities intends to remind us that they refer to the same input 1. The corresponding dual problem then becomes

Dual minimize $TC = 15y_1' - 15y_1'' + 10y_2 + 0y_3$

$$
\begin{array}{rclcrclcl}
\text{subject to} \quad 4y_1' & - & 4y_1'' & + & 2y_2 & - & y_3 & \geq & 30 \\
3y_1' & - & 3y_1'' & + & y_2 & + & 5y_3 & \geq & 45 \\
y_1' & & & & & & & \geq & 0 \\
 & & y_1'' & & & & & \geq & 0 \\
 & & & & y_2 & & & \geq & 0 \\
 & & & & & & y_3 & \geq & 0
\end{array}
$$

The same dual problem can be rewritten in a more compact form by factoring out the coefficients of the first two columns:

Dual minimize $TC = 15(y_1' - y_1'') + 10y_2 + 0y_3$

subject to
$$4(y_1' - y_1'') + 2y_2 - y_3 \geq 30$$
$$3(y_1' - y_1'') + y_2 + 5y_3 \geq 45$$
$$(y_1' - y_1'') \quad\quad\quad\quad \text{free}$$
$$y_2 \quad\quad \geq 0$$
$$y_3 \geq 0$$

The crucial information to be derived from this formulation is that the dual expression $(y_1' - y_1'')$ is free because it is the difference between two nonnegative variables. In other words, "free" means that it is not restricted in any interval of the real line and can take on any value, positive, zero, or negative. For this reason it is also called unrestricted.

Our goal is now in sight. Let us define the dual variable y_1 as $y_1 = y_1' - y_1''$. Then the correct specification of the primal and dual problems can be reformulated in final form as

Primal	**Dual**

maximize $NR = 30x_1 + 45x_2$ maximize $TC = 15y_1 + 10y_2$

subject to subject to

$4x_1 + 3x_2 = 15$	$\longleftrightarrow$	y_1 \quad\quad\quad\quad free
$2x_1 + x_2 \leq 10$	$\longleftrightarrow$	$y_2 \quad\quad \geq 0$
$-x_1 + 5x_2 \leq 0$	$\longleftrightarrow$	$y_3 \geq 0$
$x_1 \quad\quad \geq 0$	$\longleftrightarrow$	$4y_1 + 2y_2 - y_3 \geq 30$
$x_2 \geq 0$	$\longleftrightarrow$	$3y_1 + y_2 + 5y_3 \geq 45$

2.3 Dual Problems with Equality Constraints and Free Variables

We conclude that to an equality constraint there corresponds a free variable, that is, an unrestricted variable that can take on values from $-\infty$ to $+\infty$. The converse is also true: to a free variable there corresponds an equality constraint. To verify this conclusion, let us consider the following primal problem where the variable x_2 is now assumed to be free:

Primal maximize $NR = 30x_1 + 45x_2$

$$
\begin{array}{rcrcl}
\text{subject to} \quad 4x_1 &+& 3x_2 &=& 15 \\
2x_1 &+& x_2 &\leq& 10 \\
-x_1 &+& 5x_2 &\leq& 0 \\
x_1 & & &\geq& 0 \\
& & x_2 && \text{free}
\end{array}
$$

To write down the appropriate dual problem, we use all the knowledge accumulated to this point. We know, for example, that a free variable can always be represented as the difference between two nonnegative variables. Hence, let $x_2 = x_2' - x_2''$ where $x_2' \geq 0$ and $x_2'' \geq 0$. Then the primal problem can be rewritten as

Primal maximize $NR = 30x_1 + 45(x_2' - x_2'')$

dual
variables

$$
\begin{array}{rcrcll}
\text{subject to} \quad 4x_1 &+& 3(x_2' - x_2'') &\leq& 15 & \quad y_1' \\
-4x_1 &-& 3(x_2' - x_2'') &\leq& -15 & \quad y_1'' \\
2x_1 &+& (x_2' - x_2'') &\leq& 10 & \quad y_2 \\
-x_1 &+& 5(x_2' - x_2'') &\leq& 0 & \quad y_3 \\
x_1 & & &\geq& 0 & \\
& & x_2' &\geq& 0 & \\
& & x_2'' &\geq& 0 &
\end{array}
$$

This primal problem exhibits only inequalities and, therefore, we can easily write down the corresponding dual problem following the familiar rules:

Dual minimize $TC = 15(y_1' - y'') + 10y_2 + 0y_3$

$$
\begin{array}{rcrcrcl}
\text{subject to} \quad 4(y_1' - y'') &+& 2y_2 &-& y_3 &\geq& 30 \\
3(y_1' - y'') &+& y_2 &+& 5y_3 &\geq& 45 \\
-3(y_1' - y'') &-& y_2 &-& 5y_3 &\geq& -45 \\
(y_1' - y'') & & & & &\text{free}& \\
& & y_2 & & &\geq& 0 \\
& & & & y_3 &\geq& 0
\end{array}
$$

The second and third constraints represent complementary dual half spaces and can be combined into an equality constraint. Furthermore, letting $y_1 = (y_1' - y'')$, it is possible to rewrite the primal and dual problems in their final form as

	Primal			**Dual**	

maximize $NR = 30x_1 + 45x_2$ $\qquad$ minimize $TC = 15y_1 + 10y_2$

subject to $\qquad\qquad\qquad\qquad\qquad$ subject to

$$
\begin{array}{rcll}
4x_1 + 3x_2 &=& 15 \\
2x_1 + x_2 &\leq& 10 \\
-x_1 + 5x_2 &\leq& 0 \\
x_1 &\geq& 0 \\
x_2 && \text{free}
\end{array}
\qquad
\begin{array}{c}
\longleftrightarrow \\
\longleftrightarrow \\
\longleftrightarrow \\
\longleftrightarrow \\
\longleftrightarrow
\end{array}
\qquad
\begin{array}{rcll}
y_1 &&& \text{free} \\
& y_2 & \geq & 0 \\
&& y_3 \geq & 0 \\
4y_1 + 2y_2 - y_3 &\geq& 30 \\
3y_1 + y_2 + 5y_3 &=& 45
\end{array}
$$

This last formulation shows the full symmetry of linear programming in all its possible specifications. The relationships between inequalities and nonnegative variables on one side, and between equations and free variables on the other, are mathematically required but they admit a meaningful economic interpretation. A summary of the appropriate dual relations between constraints and variables is

Primal		**Dual**
maximize	$\longleftrightarrow$	minimize
$\leq$ constraint	$\longleftrightarrow$	$y \geq 0$
$x \geq 0$	$\longleftrightarrow$	$\geq$ constraint
$=$ constraint	$\longleftrightarrow$	y free
x free	$\longleftrightarrow$	$=$ constraint

Chapter 3

Setting Up LP Problems

3.1 Preliminaries

Linear programming problems are like any others: they require first a careful and complete description. Then the procedure of setting up a problem within a framework suitable for linear programming is principally a matter of common sense and simple logic. No theory of setting up linear programming problems exists. One must begin by studying a few examples, trying to grasp their essential features and then attempting to adapt these features to new problems. Although much of this book's interest rests with the ability to solve linear programming problems and to analyze the corresponding results from an economic viewpoint, one can hardly achieve these goals without the ability to set up the problems properly.

In general, the information pertaining to a problem is available in a seemingly disorderly fashion, possibly combined with irrelevant data. In order to sort out the relevant information and relations, the following hints are of great help:

1. Read the description of the problem very carefully while identifying by name the activities of the problem. An *activity* is a way of doing one thing; it is a process. It describes in detail the necessary steps and input quantities required to produce a unit of one or more outputs. An activity, therefore, is a blueprint of technology. For example, if corn can be produced with two different quantities of labor and machinery, there are two activities for the same corn commodity. One activity can also represent the production of two (or more) joint products such as wheat grain and wheat straw.

2. Identify by name the *constraints* of the problem. In general, these constraints are in the form of either input or output limitations.

3. Identify the measurement units of both activities and constraints.

4. Identify the technical coefficients and their measurement units relating an activity to a constraint.

5. Using words at first, set up the appropriate relations (inequalities, equations) among activities and constraints. Given the economic context used in this book, the following general relation is of great help in formulating constraint relations in a correct way:

$$\text{consumption} + \text{investment} \leq \text{production} + \text{availability}$$

The above relation concerns any commodity, whether input or output. Often, when dealing with inputs like land, labor, etc., the "production" terminology is changed into "availability" (commodity already produced). Furthermore, the "consumption" category may not be of interest in many problems and, therefore, is left out. Finally, the "investment" category must be broadly understood as investing in a technology for the purpose of producing commodities at an unknown level. Hence, alternative formulations of the fundamental relation are

$$\text{technology} \times \text{unknown level of commodities} \leq \text{resource availability}$$
$$\text{investment} \leq \text{production}$$
$$\text{demand} \leq \text{supply}$$

6. After all the constraint relations are specified according to the above suggestions, a symbolic notation is assigned to the unknown activity levels, while the terms in the constraints are rearranged to have all the variables on one side and the constant terms on the other side of the inequality (equation).

7. Finally, lift the variable names from the constraint relations and transfer the remaining numerical coefficients into an initial tableau.

The procedure outlined above will be illustrated with several examples.

3.2   Problem 1: A Small Family Farm

Description. A farmer owns 10 acres of land that he wishes to cultivate with wheat and alfalfa. He also owns a small cow shed capable of housing 15 dairy cows. It takes .1 acre to produce 10 bushels of wheat and half a ton of straw. The production of 1 ton of alfalfa requires .05 acre of land (irrigated and multiple cuts per year). Alfalfa and straw are entirely destined to be used by the dairy cows. During a year, every cow produces 20,000 pounds of milk and requires 15 tons of alfalfa and 3 tons of straw. Wheat can be sold at $5 per bushel and milk at $.15 per pound. Of course, the farmer wants to maximize sales revenue.

The task is to interpret the description of the problem, to set up the corresponding primal linear programming specification, to show all the problem's coefficients in a tableau, to write down the corresponding dual problem, and finally, to give an economic interpretation of each component, including the objective function and the constraints.

The strategy for setting up a linear programming problem is the one already outlined:

1. Identify the activities and give them a name.
2. Identify the constraints and give them a name.
3. Relate activities to constraints via the fundamental relation

$$\text{demand} \leq \text{supply}$$

Solution. The activities identified in this problem are wheat, alfalfa, and milk production. The wheat activity is an example of joint production, since the final outputs obtained from .1 of an acre are 10 bushels of grain and half a ton of straw. It is convenient to list the identified activities together with the proper measurement units as follows:

Activities:
W ≡ wheat measured in units of 10 bushels of grain
A ≡ alfalfa measured in units of 1 ton of hay
C ≡ milk measured in units of 1 cow

It is clear, then, that there is considerable freedom in the choice of measurement units. Indeed, activities can be measured in terms of either output or input units. The final choice is left either to convention or to convenience. In this example, wheat and alfalfa are measured in output units (10 bushels of grain and 1 ton of hay) while milk is measured in input units (cows). It is important to emphasize from the start that the choice of units could have been different. The wheat and alfalfa activities could have been measured in units of land (acres). Similarly, the milk activity could have been measured in units of 20,000 pounds of milk.

The identification of the limiting requirements results in the following:

Constraints:
1. ≡ land measured in acres
2. ≡ cow shed measured in cow units
3. ≡ alfalfa availability plus production measured in tons
4. ≡ straw availability plus production measured in tons

Both straw and alfalfa must be produced, since initial availabilities are nonexistent. Hence, these two commodities must be considered as outputs and inputs in the same production plan.

The specification of the relation among activities and constraints is the next step. In this part of the solution we must specify one inequality (or equality, if necessary) relation for each constraint identified above:

1. wheat land + alfalfa land $\leq$ available land
$$.1W + \qquad .05A \leq 10$$
2. space required by dairy cows $\leq$ space in cow shed
$$C \leq 15$$
3. alfalfa required by dairy cows $\leq$ alfalfa produced
$$15C \leq A$$
4. straw required by dairy cows $\leq$ straw produced by wheat activity
$$3C \leq .5W$$

Let us verify the question involving the measurement units by analyzing the first constraint, rewritten here in a more explicit way as

$$\frac{\text{acres}}{10 \text{ bu } W} \times (\text{ no. of 10 bu } W) + \frac{\text{acres}}{\text{ton of } A} \times (\text{ no. of tons of } A) \leq \text{ no. of acres}$$

The coefficient of the wheat activity, .1, is measured in acres per 10 bushels of grain, and the coefficient of the alfalfa activity is measured in acres per 1 ton of hay. The coherence of the specification in terms of measurement units is, therefore, verified by noticing that, in each term, the denominator is cancelled with a corresponding numerator, leaving each addendum measured in terms of acres. Each constraint must satisfy the same consistency test: the terms on the left-hand side of the inequality must be measured in the same units of the term on the right-hand side.

As another example we analyze the fourth constraint, the straw constraint:

$$\frac{\text{tons of straw}}{\text{cow}} \times (\text{no. of cows}) \leq \frac{\text{tons of straw}}{10 \text{ bu of } W} \times (\text{no. of 10 bu of } W)$$

Again, the cancellation between numerators and denominators explicitly leaves the left-hand side measured in the same units of the right-hand side.

It remains to specify the proper objective function, remembering that straw and alfalfa are not sold directly on the market. Hence, the primal objective function is

$$\text{maximize revenue} = \text{revenue from wheat grain} + \text{revenue from milk}$$
$$= \$(5 \times 10)W + \$(.15 \times 20,000)C$$

Finally, the specification of the primal problem can be stated in compact form as follows:

Primal maximize $R = 50W + 3000C$

				dual variables
subject to	$.1W$	$+ .05A \leq 10$		y_1
	C	≤ 15		y_2
	$15C -$	$A \leq 0$		y_3
	$-.5W + 3C$	≤ 0		y_4

$$W, A, \text{ and } C \geq 0$$

The negative signs associated with the A and W activities in the third and fourth constraint are the consequence of rearranging all the variables on the left-hand side of the inequalities, as recommended in rule 6 of section 3.1. Notice also how the joint production of wheat grain and straw affects the specification of the wheat activity: wheat (W) produces 10 bushels of grain that are sold directly on the market and, therefore, appear in the generation of the unit revenue for wheat in the objective function; the straw produced by the wheat activity is not sold on the market but is used for raising the dairy cows and, thus, it shows up in the straw constraint.

The initial tableau. The orderly solution of a linear programming problem is conveniently accomplished by arranging all the coefficients into an array, called a tableau. Although this method will be explained in detail in the next few chapters, it is useful to practice this operation from the start. The initial tableau is constructed by adding all the appropriate slack variables to the constraints of the problem and then lifting the variable names from all the equations, including the objective function. The names of the variables appear as headings for the corresponding columns of coefficients. All the equality signs belonging both to the constraints as well as to the objective function have been aligned in front of the solution column, a required operation in order to keep the column alignment among all the variables. This is also the reason for the negative signs associated with the coefficients of the objective function. Those coefficients are separated by a dotted line at the bottom of the tableau, as a reminder of their special nature. They are placed at the bottom of the tableau for no special reason. Alternatively, they could have been placed at the top of it without affecting the significance of the arrangement. The word "solution" (abbreviated sol) above the constraint column indicates the place where the intermediate and final solutions will appear during computations.

Initial Tableau

$$
\begin{bmatrix}
R & \vdots & W & C & A & x_{s1} & x_{s2} & x_{s3} & x_{s4} & \vdots & & \text{sol} \\
0 & \vdots & .1 & 0 & .05 & 1 & 0 & 0 & 0 & \vdots & = & 10 \\
0 & \vdots & 0 & 1 & 0 & 0 & 1 & 0 & 0 & \vdots & = & 15 \\
0 & \vdots & 0 & 15 & -1 & 0 & 0 & 1 & 0 & \vdots & = & 0 \\
0 & \vdots & -.5 & 3 & 0 & 0 & 0 & 0 & 1 & \vdots & = & 0 \\
\cdots & & & & & & & & & & & \\
1 & \vdots & -50 & -3000 & 0 & 0 & 0 & 0 & 0 & \vdots & = & 0
\end{bmatrix}
$$

The dual problem. The dual problem can now be formulated, keeping in mind the transposition rules identified in section 1.8.

Dual minimize $TC = 10y_1 + 15y_2$

$$
\begin{aligned}
\text{subject to} \quad .1y_1 \qquad\qquad - .5y_4 &\geq 50 \\
y_2 + 15y_3 + 3y_4 &\geq 3000 \\
.05y_1 \qquad - y_3 \qquad &\geq 0
\end{aligned}
$$

$$\text{all } y_i \geq 0, \quad i = 1, \ldots, 4$$

As can be easily verified, the dual problem utilizes the information contained in the initial tableau under the headings of the wheat (W), milk (C), and alfalfa (A) activities as well as under the solution column. The other columns belonging to the slack variables do not contain much information and are summarized in the nonnegative dual variables.

Interpretation of the dual problem. As indicated in chapter 1, the dual variables have the meaning of imputed prices corresponding to limiting inputs. Often, the same variables are also interpreted as *shadow* prices, in contrast to *market* prices, which are visible to everybody. Another expression for the dual variables is the marginal sacrifice that an economic agent must bear because of the presence of a constraint. If that constraint could be made less limiting by one unit, the value of the objective function would increase by the amount given by the dual variable. Conversely, if the constraint becomes tighter by one unit, the objective function will decrease by the value of the corresponding dual variable. In either case, we think of a dual variable in terms of a marginal sacrifice (a profit is a negative sacrifice) that must be borne because the corresponding constraint is binding or, in other words, it is satisfied with the equality sign.

With this general interpretation of the dual variables we can proceed to analyze the dual problem as a whole and in its various components.

The farmer, of course, wishes to minimize the imputed cost of the limiting resources (land and cow shed) that make up the farm subject to the equilibrium conditions expressed by the requirement that, for each line of production, the marginal cost is greater than or equal to the corresponding marginal revenue. An important economic lesson to be derived from studying the dual problem is that it is not convenient for the farmer to overvalue the farm. In other words, why should a farmer minimize (rather than maximize) the value of the land? Because maximizing the value of productive resources is not an equilibrium condition. If the unit value of land, the shadow price of it, is too high, the marginal cost of those products utilizing that resource will also be too high and will erroneously indicate to the farmer that producing those outputs is not profitable. Other farmers will soon realize what is going on with their neighbor and will enter into a bid to acquire the farm. They clearly wish to pay the minimum amount of dollars for that farm and, therefore, they will offer prices for the land and the cow shed that will minimize the total cost of buying the farm. This is the fundamental reason why it is not convenient for the farm operator to overvalue it.

The economic interpretation of the dual constraints is based upon the general scheme that $MC \geq MR$, but it is instructive to be able to apply this framework to each individual constraint.

Constraint 1: $.1y_1 \geq 50 + .5y_4$

For an easier and correct economic interpretation of any constraint it is important to rearrange the various terms and eliminate all the negative signs from the relation. The dual variable y_4 is the value of each unit of straw destined for the dairy cows. Hence, the marginal cost of the wheat activity that utilizes only land $(.1y_1)$ must be greater than or equal to the marginal revenue of wheat grain (50) plus the marginal value product of the wheat activity in terms of straw $(.5y_4)$. In other words, revenue from the wheat activity is made up of direct revenue from wheat grain plus the indirect revenue from wheat straw utilized for producing milk.

Constraint 2: $y_2 + 15y_3 + 3y_4 \geq 3000$

The marginal cost of milk (measured in cow units) is given by the left-hand side of the constraint, since y_2 is the marginal cost of one stall in the cow shed, $15y_3$ is the marginal cost of alfalfa, and $3y_4$ is the marginal cost of straw. The combined marginal cost of milk, then, must be greater than or equal to the marginal revenue of milk derived from one cow (3,000).

Constraint 3: $.05y_1 \geq y_3$

The dual variable y_3 is the imputed value of each unit of alfalfa destined for the dairy cows. Hence, the marginal cost of land destined for alfalfa production must be at least as large as the value marginal product of alfalfa.

The process of obtaining a numerical solution for a linear programming problem will be discussed at length in subsequent chapters. Nonetheless, we can present here the optimal primal and dual solutions of this example as an illustration of what kind of information can be expected from a linear programming problem.

The optimal production plan recommends allocating the available 10 acres of land between wheat and alfalfa for the production of 444.444 bushels of grain, 22.222 tons of straw, and 111.111 tons of hay. Given the input land coefficients for wheat (.1) and alfalfa (.05), the allocation suggests that 4.4444 (.1 × 44.444) acres of land be destined for wheat and 5.5556 (.05 × 111.111) acres for alfalfa. With those amounts of alfalfa and straw producible with this plan, it is possible to feed and raise 7.4074 cows (please, downplay the fractional cow) with 7.5926 unused stalls in the cow shed. The level of revenue achievable with this production plan is given by the primal objective function: ($24, 444.4 = $50 × 44.444 + $3, 000 × 7.4074).

The optimal dual solution indicates that the shadow price of an acre of land is $2,444.44, the shadow price of a ton of alfalfa hay is $122.22, and that of a ton of straw is $388.89. The shadow price of a stall in the cow shed is equal to $0 because some of the initially available stalls remain vacant under the optimal plan. The level of the dual objective function is $24, 444.44 = $2, 444.44 × 10. Notice that the levels of the primal and dual objective functions are the same, although they are computed with two different sets of numbers. As additional information it is possible to obtain the cost of alfalfa ($122.222 × 111.111 = $13, 580.21) and that of straw ($388.889 × 22.222 = $8, 641.89). The total cost of raising the number of cows specified by the plan is $22, 222.1 = $13, 580.21 + $8, 641.89, which is equal (except for rounding errors) to the revenue obtained from selling the milk produced by those cows ($22, 222.22 = $3, 000 × 7.4074). Even as simple a problem as that presented in this example has turned out to be rather informative.

3.3   Problem 2: The Expanded Family Farm

The following problem is an extension of problem 1. To facilitate its reading we report the description of it with the additional information in *italics*.

Description. A farmer owns 10 acres of land that he wishes to cultivate with wheat and alfalfa. He also owns a small cow shed capable of housing 15 dairy cows. It takes .1 acre to produce 10 bushels of wheat and half a ton of straw. The production of 1 ton of alfalfa requires .05 acre of land (irrigated and multiple cuts per year). Alfalfa and straw *either can be sold on the market for $120 and $30 per ton, respectively, or can be* destined to the dairy cows. During a year, every cow produces 20,000 pounds of milk requiring 15 tons of alfalfa and 3 tons of straw. Wheat can be sold at $5.00 per bushel and milk at $.15 per pound. Of course, the farmer wants to maximize sales revenue.

The task is, again, to interpret the description of the problem, to set up the corresponding primal linear programming specification, to display all the problem's coefficients in a tableau, to write down the corresponding dual problem, and finally, to give an economic interpretation of each component, including the objective function and the constraints.

Solution. We proceed, as usual, with the routine but crucial step of identifying all activities and constraints.

Activities: In problem 1 there were three activities (wheat, alfalfa, and milk) that produced four commodities (wheat grain, wheat straw, alfalfa hay, and milk). In problem 2 alfalfa and straw can be either marketed or used for raising dairy cows. Four new activities, therefore, must be added to the old list:

W $\equiv$ wheat measured in units of 10 bushels of grain
A $\equiv$ alfalfa produced measured in units of 1 ton of hay
MA $\equiv$ marketed alfalfa measured in units of 1 ton of hay
CA $\equiv$ alfalfa fed to cows measured in units of 1 ton of hay
MS $\equiv$ marketed straw measured in units of 1 ton
CS $\equiv$ straw destined to dairy cows measured in units of 1 ton
C $\equiv$ milk measured in units of 1 cow

Constraints: The alfalfa and straw constraints of problem 1 must be modified to accommodate the option of marketing those commodities. Hence, problem 2 requires two additional constraints that can be called the alfalfa and straw allocations and are specified as

marketed alfalfa   +   alfalfa fed to cows $\leq$ alfalfa produced

$$MA \quad + \quad CA \quad \leq \quad A$$

A similar constraint is required to represent the straw allocation:

marketed straw   +   straw used for cows $\leq$ straw produced

$$MS \quad + \quad CS \quad \leq \quad .5W$$

The primal objective function must also be expanded to account for the new marketing activities concerning alfalfa and straw. Hence, its detailed specification is

maximize $R = \$(5 \times 10)W + \$(.15 \times 20,000)C + \$120MA + \$30MS$

The primal problem can now be stated in its entirety as

Primal   maximize $R = 50W + 3000C + 120MA + 30MS$

subject to   dual
                                                                variables

land	$.1W$		$+ .05A$					≤ 10	y_1
cow shed		C						≤ 15	y_2
alfalfa allocation			$-\ A + MA + CA$					≤ 0	y_3
straw allocation	$-.5W$					$+ MS + CS$		≤ 0	y_4
alfalfa for cows		$15C$			$- CA$			≤ 0	y_5
straw for cows		$3C$					$- CS$	≤ 0	y_6

$$W,\ C,\ A,\ MA,\ CA,\ MS,\ \text{and } CS \geq 0$$

Initial Tableau

R	W	C	A	MA	CA	MS	CS	slack variables	sol
⋮	0.1		0.05			1		⋮	$= 10$
⋮		1					1	⋮	$= 15$
⋮			-1	1	1			1 ⋮	$= 0$
⋮	-0.5				1	1		1 ⋮	$= 0$
⋮		15		-1				1 ⋮	$= 0$
⋮		3				-1		1 ⋮	$= 0$
1 ⋮	-50	-3000	0	-120	0	-30		0 0 0 0 0 0 ⋮	$= 0$

In the above initial tableau the zero coefficients have been left out to allow
a better vision of the relations among variables and constraints.
    The dual problem is as follows:

Dual minimize $TC = 10y_1 + 15y_2$

subject to	$.1y_1$			$-\ .5y_4$			$\geq$	50
		y_2			$+\ 15y_5 + 3y_6$		$\geq$	3000
	$.05y_1$		$-\ y_3$				$\geq$	0
			y_3				$\geq$	120
			y_3		$-\ y_5$		$\geq$	0
				y_4			$\geq$	30
				y_4		$-\ y_6$	$\geq$	0

$$\text{all } y_i \geq 0, \quad i = 1, \ldots, 6$$

Economic interpretation. Although some of the constraints have the same structure and meaning as those in the dual of problem 1, it is convenient to reinterpret the entire set of dual relations in an orderly fashion because some dual variables have been renumbered.

Constraint 1: $.1y_1 \geq 50 + .5y_4$

The wheat activity utilizes only land. Hence, the quantity $(.1y_1)$ is the marginal cost of the wheat activity, since the coefficient .1 specifies the land requirement and y_1 is the shadow price of land. The first dual constraint, therefore, stipulates that the marginal cost of wheat must be greater than or equal to the marginal revenue of wheat grain (50) augmented by the value marginal product of straw $(.5y_4)$.

Constraint 2: $y_2 + 15y_5 + 3y_6 \geq 3000$

The marginal cost of milk (measured in cow units) must be greater than or equal to the marginal revenue produced, on the average, by a cow. The dual variable y_2 is the imputed price of a stall in the cow shed, y_5 is the shadow price of alfalfa fed to cows, and y_6 is the imputed price of straw used for the dairy cows.

Constraint 3: $.05y_1 \geq y_3$

The marginal cost of alfalfa that uses only land $(.05y_1)$ must be greater than or equal to the value marginal product of alfalfa (y_3).

Constraint 4: $y_3 \geq 120$

The shadow price of the total alfalfa produced must be greater than or equal to 120, the marginal revenue of alfalfa sold on the market. Alfalfa is both an input and an output.

Constraint 5: $y_3 \geq y_5$

The value marginal product of total alfalfa produced must also be greater than or equal to the value marginal product of that portion of alfalfa fed to cows.

Constraint 6: $y_3 \geq 30$

The value marginal product of total straw produced must be greater than or equal to the marginal revenue of the portion of straw sold on the market. As with alfalfa, straw is both an input and an output.

Constraint 7: $y_4 \geq y_6$

The value marginal product of total straw produced must be greater than or equal to that of the straw used for cows.

It is of interest to notice that the primal of problem 2 can have an alternative and equivalent formulation obtained by adding constraints 3 and 5 as well as 4 and 6. In this way, variables CA and CS, the amount of alfalfa and straw destined for cows, apparently disappear from the specification but, in reality, there is no loss of information. The activities associated with the variables CA and CS are called *transfer* activities in LP terminology because they transfer the information from one constraint to another. They are a very useful type of activity whenever one wishes to obtain an explicit accounting of the solution results.

With the addition of the constraints indicated above, the new equivalent version of the primal problem is

Primal   maximize $R = 50W + 3000C + 120MA + 30MS$

subject to

					dual variables
land	$.1W$	$+ .05A$		≤ 10	y_1
cow shed		C		≤ 15	y_2
alfalfa allocation		$15C - A + MA$		≤ 0	y_3
straw allocation	$- .5W +$	$3C$	$+ MS$	≤ 0	y_4

$$W, A, C, MA, \text{ and } MS \geq 0$$

In this more compact version of problem 2, the information pertaining to the amount of alfalfa and straw allocated to cows is implicitly defined by the quantities $15C$ and $3C$, which represent the requirements of alfalfa and straw necessary to raise those cows.

The numerical solution of this expanded family farm problem is identical to that of the previous example. The virtue of this formulation lies in the introduction of transfer activities that connect the production process to the distribution channels of marketing and internal consumption.

3.4   Problem 3: A Grape Farm

Description. A farmer owns 100 acres of Thompson seedless grapes. These grapes have three kinds of utilization: as table grapes, raisins, and wine grapes. The decision about the preferred utilization is made yearly by the farmer who, therefore, can take advantage of market conditions.

An acre of land produces 30 tons of grapes. If they are sold as table grapes, the farmer must package them and can receive $300 per ton. It takes two workdays of eight hours to package a ton of grapes. Workers can be hired for $4 an hour. If sold as raisins, grapes must be dried. A ton of fresh grapes produces .2 ton of raisins. It takes four workdays to completely dry and package a ton of raisins. The farmer can receive \$800 per ton. If sold as wine grapes, the farmer can receive \$200 per ton. It takes only .3 workday per ton to ready wine grapes for delivery to a winery. The farmer wishes to maximize net revenue.

As usual, we must set up the primal LP problem, its corresponding dual, and provide a meaningful economic interpretation of all its constraints.

Solution. In this section we identify all the activities and constraints.

Activities:

$RG \equiv$ raisin grapes measured in tons
$TG \equiv$ table grapes measured in tons
$R \equiv$ raisins measured in tons
$WG \equiv$ wine grapes measured in tons
$L \equiv$ hired labor measured in 8-hour workdays

Constraints:

$G \equiv$ total grapes (land $\times$ yield) measured in tons
$P \equiv$ proportion of raisins to grapes
$L \equiv$ labor measured in workdays

Primal

maximize $NR = 300TG + 200WG + 800R - 32L$

$$
\begin{aligned}
\text{subject to} \quad TG + RG + WG &\leq 3000 \\
R &= .2RG \\
2TG + .3WG + 4R &\leq L \\
TG,\ RG,\ WG,\ R,\ \text{and}\ L &\geq 0
\end{aligned}
$$

The objective function reflects the possibility of hiring labor by means of a labor hiring activity. It represents a cost and, therefore, it appears with a minus sign. For this reason the proper meaning of the objective function is that of net revenue, since labor costs are netted out of revenue. The first constraint represents total grapes produced. The coefficients of 3,000 tons of grapes is obtained as a product of 100 acres of land per 30 tons of yield. The second constraint is novel in our empirical formulations and represents the shrinkage of raisin grapes in the production of raisins. It must be stated as an equality following the given information and, therefore, its corresponding dual variable will be a free variable. Accordingly, no slack variable is admissible for this constraint. The third constraint represents the labor requirements and availability. The second and the third constraints exhibit variable quantities on the right-hand side because the proper procedure is to initially set up the primal problem following the fundamental relation (demand $\leq$ supply) with no need to worry about quantities associated with negative signs. After the constraints are correctly specified, those variable quantities are brought to the left-hand side as in the following initial tableau:

Initial Tableau

NR		TG	RG	WG	R	L	x_{s1}	x_{x3}		sol
	⋮	1	1	1		1		⋮	=	3000
	⋮		−.2		1			⋮	=	0
	⋮	2		.3	4	−1		1 ⋮	=	0
1	⋮	−300	0	−200	−800	32	0	0 ⋮	=	0

The dual problem takes on the following structure:

Dual minimize $TC = 3000y_1$

$$
\begin{array}{lllll}
\text{subject to} & y_1 & + 2y_3 \geq 300 & \text{(A)} \\
& y_1 - .2y_2 & \geq 0 & \text{(B)} \\
& y_1 & + .3y_3 \geq 200 & \text{(C)} \\
& y_2 + 4y_3 \geq 800 & & \text{(D)} \\
& - y_3 \geq -32 & & \text{(E)}
\end{array}
$$

y_1 and $y_3 \geq 0$, y_2 free

Economic interpretation. In this simplified problem, the essentially limiting input is produced grapes whose total value must be minimized, as indicated by the objective function. Again, notice the correspondence between the coefficients of the initial tableau and those of the dual problem. Each row of coefficients in the tableau (excluding those pertaining to the slack variables) appears as a column in the dual problem.

Constraint (A) states that the marginal activity cost of 1 ton of table grapes ($y_1 + 2y_3$) must be greater than or equal to $300, the marginal revenue of table grapes. The variable y_1 is the shadow price of fresh grapes and $2y_3$ is the marginal cost of labor used in the table grapes activity. To interpret constraint (B), it is convenient to restate it as $y_1 \geq .2y_2$, indicating that the value marginal product of fresh grapes must be greater than or equal to one-fifth of the value marginal product of raisins. Constraint (C) states that the marginal cost of wine grapes must be greater than or equal to $200, the marginal revenue of wine grapes. Similarly, constraint (D) asserts the same relationship for raisins. Finally, constraint (E), rewritten as $y_3 \leq 32$, requires that the shadow price of labor measured in workdays must be less than or equal to its market price of $32.

The numerical solution of this problem recommends the sale of all available 3,000 tons of grapes as table grapes with a requirement of 6,000 workdays of hired labor. The maximum level of revenue attainable is $708,000 = $300 × 3,000 − $32 × 6,000. The dual solution shows that a ton of fresh grapes is valued at $236, while the value marginal product of a workday of labor is $32, the same as the market price for hired labor. The optimal dual objective function is $708,000 = $236 × 3,000, the same as the primal counterpart.

3.5   Problem 4: The Expanded Grape Farm

This problem is similar to that described in problem 3. Additional institutional and subjective information is introduced (in *italics*) to illustrate the flexibility of linear programming.

Description. A farmer owns 100 acres of Thompson seedless grapes. These grapes have three kinds of utilizations: as table grapes, raisins, and wine grapes. The decision about the preferred utilization is made yearly by the farmer who, therefore, can take advantage of market conditions.

An acre of land produces 30 tons of grapes. If they are sold as table grapes, the farmer must package them and can receive $300 per ton. It takes two workdays of eight hours to package a ton of grapes. *During the harvest period the farmer has at his disposal 135 workdays of family labor.* Furthermore, other workers can be hired for $4 an hour. If sold as raisins, grapes must be dried. A ton of fresh grapes produces *between .18 and .2* ton of raisins. It takes four workdays to completely dry and package a ton of raisins. The farmer can receive $800 per ton. If sold as wine grapes, the farmer can receive $200 per ton. It takes only .3 workday per ton to ready wine grapes for delivery to a winery. *In past years, the farmer observed that a profitable proportion of table grapes varied between one-fourth and one-half of wine grapes. He thinks that this information should be part of his plan.* The farmer, of course, wishes to maximize net revenue.

We must set up the primal problem, the initial tableau, and the corresponding dual problem and provide an economic interpretation of it.

Solution

Activities: The activities are the same as in problem 3.

RG	≡	raisin grapes measured in tons
TG	≡	table grapes measured in tons
R	≡	raisins measured in tons
WG	≡	wine grapes measured in tons
L	≡	hired labor measured in 8-hour workdays

Constraints: The constraints are those of problem 3 plus three additional ones dealing with ratios of table grapes to raisins and wine grapes.

G ≡ total grapes (land × yield) measured in tons
$P1$ ≡ upper proportion of raisins to grapes
$P2$ ≡ lower proportion of raisins to grapes
L ≡ labor measured in workdays
$R1$ ≡ upper proportion of table grapes to wine grapes
$R2$ ≡ lower proportion of table grapes to wine grapes

Primal

maximize $NR = 300TG + 200WG + 800R - 32L$

$$
\begin{array}{lrcl}
 & & & \text{dual} \\
 & & & \text{variables} \\
\text{subject to} \quad TG + RG + \;\; WG & \leq 3000 & & y_1 \\
R \leq .2RG & & & y_2' \\
R \geq .18RG & & & y_2'' \\
2TG \qquad + .3WG + 4R \leq L + 135 & & & y_3 \\
TG \qquad\qquad \leq .5WG & & & y_4' \\
TG \qquad\qquad \geq .25WG & & & y_4''
\end{array}
$$

$TG,\; RG,\; WG,\; R,$ and $L \geq 0$

Dual variables pertaining to the same type of constraints have been assigned the same subscript. Notice also that, in this example, we have stated the inequality constraints in either direction as a more natural way to read the problem. In this way, no negative signs appear in the primal specification. However, the correct formulation of the dual problem requires conceiving the inequalities as running in the same direction.

Initial Tableau

NR		TG	RG	WG	R	L		sol	
	⋮	1	1	1			⋮	≤	3000
	⋮		−.2		1		⋮	≤	0
	⋮		.18		−1		⋮	≤	0
	⋮	2		.3	4	−1	⋮	≤	135
	⋮	1		−.5			⋮	≤	0
	⋮	−1		.25			⋮	≤	0
1	⋮	−300	0	−200	−800	32	⋮	=	0

In the above initial tableau the slack variables are purposely left out in order to train the reader to recognize their presence by means of the inequality sign associated with the corresponding constraints.

Dual minimize $TC = 3000y_1 + 135y_3$

$$
\begin{aligned}
\text{subject to } y_1 && + 2y_3 + y_4' - y_4'' &\geq 300 \\
y_1 - .2y_2' + .18y_2'' && &\geq 0 \\
y_1 && + .3y_3 - .5y_4' + .25y_4'' &\geq 200 \\
y_2' - y_2'' + 4y_3 && &\geq 800 \\
- y_3 && &\geq -32
\end{aligned}
$$

$$\text{all } y_i \geq 0, \quad i = 1, \ldots, 4$$

The economic interpretation of this dual problem is very similar to that of problem 3, with only minor modifications. Such modifications, however, are important for comprehending the structure of marginal cost and marginal revenue associated with each activity. For this reason we will restate below the dual constraints in a manner that clearly reflects the relationship between marginal cost (MC) and marginal revenue (MR).

$$
\begin{array}{llcccc}
& & MC & \geq & MR & \\
\text{table grapes } y_1 & & + 2y_3 + y_4' & \geq & 300 + & y_4'' \\
\text{raisin grapes } y_1 & + .18y_2'' & & \geq & .2y_2' & \\
\text{wine grapes } y_1 & & + .3y_3 + .25y_4'' & \geq & 200 + & .5y_4' \\
\text{raisins } & y_2' & + 4y_3 & \geq & 800 + & y_2''
\end{array}
$$

With the additional constraints imposed on the desirable proportion of wine to table grapes, the numerical solution of this expanded grape problem indicates that the optimal production plan is composed of 1,000 tons of table grapes and 2,000 tons of wine grapes. The combined types of grapes require 2,465 workdays of hired labor. The optimal level of revenue is $621,120, down from the $708,000 of the preceding example because of the additional constraints.

The dual solution shows that a ton of fresh grapes at the farm is valued at $205.6, while labor's shadow price is still $32 per workday. The minimum total cost of the production plan is given by the dual objective function and amounts to ($621,120 = $205.6 × 3,000 + $32 × 135), which is, again, equal to the level of the primal objective function.

3.6   Problem 5: A Sequential Production

Description. A family farmer, with three family workers at his disposal all year, owns 100 acres of land and may rent additional land at $200 per acre. He considers the formulation of a production plan that includes wheat, tomatoes, alfalfa, and milk. Wheat and tomatoes, if produced, are grown in sequence (rotation) on the same land; wheat is planted first and the tomatoes will follow it, although not all the wheat acreage would necessarily be farmed with tomatoes.

There is also a 150-cow feedlot. Each cow eats 15 tons of alfalfa hay per year and produces 20,000 pounds of milk valued at $.15 per pound. Each cow requires .05 work-year of labor. An acre of wheat gives 22,000 pounds of grain saleable at $.10 per pound and requiring .15 work-year of labor. An acre of tomatoes produces 90,000 pounds of fruit that can be sold for $.05 per pound, requiring .3 work-year of labor. An acre of alfalfa can produce 40,000 pounds of hay, requiring .2 work-year of labor. Workers can also be hired at $12,000 per year. The farmer wishes to maximize net revenue.

This problem is similar to others already discussed at length. The additional feature is the introduction of the idea of rotation, that is, a series of production activities linked in time through the same land. Notice that the crop that follows a previous crop (tomatoes follow wheat, in this example) does not need to be planted on the entire acreage. This specification adds flexibility to the production plan and results in an inequality (rather than an equality) constraint. An important recommendation for setting up linear programming models, therefore, is to avoid as much as possible the formulation of constraints in equality form, since they may limit the possibility of achieving a feasible solution.

Solution

Activities:

$W \equiv$ wheat
$T \equiv$ tomatoes
$A \equiv$ alfalfa
$C \equiv$ milk
$R \equiv$ land renting
$L \equiv$ labor hiring

Constraints:

$1 \equiv$ land
$2 \equiv$ cow shed
$3 \equiv$ wheat-tomatoes land
$4 \equiv$ alfalfa fed to cows
$5 \equiv$ labor

Primal

maximize $NR = \$(.10 \times 22,000)W + \$(.05 \times 90,000)T$
$+\$(.15 \times 20,000)C - \$200R - \$12,000L$

subject to

$$W + A \leq 100 + R$$
$$C \leq 150$$
$$T \leq W$$
$$15C \leq 20A$$
$$.15W + .3T + .2A + .05C \leq 3 + L$$

$$W, T, A, C, R, \text{ and } L \geq 0$$

In setting up the above system of constraints we have kept closely in mind the structure of the fundamental relation: demand $\leq$ supply. Two main points must be emphasized. The tomato activity (T) does not appear in the land constraint because it will use the wheat land after harvest. The introduction of T into the land constraint would destroy the notion of rotation and cause a misspecification of the problem. Secondly, attention must always be paid to the units of measurement for activities and coefficients. For example, the coefficient (20) associated with alfalfa activity in the fourth constraint is the result of transforming alfalfa production, reported as 40,000 pounds per acre, into tons per acre. This transformation is necessary because the feeding requirements of cows is given in tons per cow.

Initial Tableau

NR		W	T	A	C	R	L		sol	
	:	1		1		−1		:	≤	100
	:				1			:	≤	150
	:	−1	1					:	≤	0
	:			−20	15			:	≤	0
	:	.15	.3	.2	.05		−1	:	≤	3
1	:	−2200	−4500	0	−3000	200	12,000	:	=	0

Dual

minimize $TC = 100y_1 + 150y_2 + 3y_5$

$$
\begin{aligned}
\text{subject to} \quad & y_1 & - y_3 & & & + .15y_5 & \geq & & 2200 & \quad \text{(A)} \\
& & y_3 & & & + .3y_5 & \geq & & 4500 & \quad \text{(B)} \\
& y_1 & & - 20y_4 & + & .2y_5 & \geq & & 0 & \quad \text{(C)} \\
& & y_2 & + 15y_4 & + & .05y_5 & \geq & & 3000 & \quad \text{(D)} \\
& -y_1 & & & & & \geq & & -200 & \quad \text{(E)} \\
& & & & - & y_5 & \geq & & -12,000 & \quad \text{(F)}
\end{aligned}
$$

$$\text{all } y_i \geq 0, \quad i = 1, \ldots, 5$$

Economic interpretation

Constraint (A): $y_1 + .15y_5 \geq 2200 + y_3$. The marginal cost of wheat must be greater than or equal to the marginal revenue of wheat plus the value marginal product of tomatoes. This added marginal revenue is due to the use of wheat land for tomatoes also.

Constraint (B): The marginal cost of tomatoes must be greater than or equal to their marginal revenue.

Constraint (C): $y_1 + .2y_5 \geq 20y_4$. The marginal cost of alfalfa, which uses land and labor, must be greater than or equal to its shadow marginal revenue, which is computed as the product of 20 tons of hay times the alfalfa shadow price, y_4.

Constraint (D): The marginal cost of milk must be greater than or equal to its marginal revenue.

Constraint (E): The shadow price of land must be less than or equal to its market price.

Constraint (F): The shadow price of labor must be less than or equal to its market price.

The dual objective function minimizes the imputed total cost of the limiting resources, land, cow shed, and family labor.

3.7   Problem 6: A Horticultural Farm

Description. A farmer owns 500 acres of arable land and wishes to formulate a production plan that includes wheat, tomatoes, cantaloupe, and cauliflower. The production season (year) is divided into three periods: October–February, March–May, and June–September. Furthermore, in each period, the farmer may rent additional land for $200 per acre up to a maximum of 100 acres or may lease out unused land for the same amount of dollars. Land rented in each period will be rented for the rest of the year.

Wheat, if produced, will occupy the land between October and May, with a possible revenue of $500 dollars per acre. Cantaloupe will occupy the land between June and September, with a revenue of $1,300 per acre. Cauliflower will occupy the land between October and February, with a revenue of $2,000 per acre. Tomatoes will occupy the land between March and September, with a revenue of $1,500 per acre. The farmer wishes to maximize net revenue.

Solution

Activities:

W ≡ wheat
T ≡ tomatoes
CT ≡ cantaloupe
CL ≡ cauliflower
RL_k ≡ renting of land in period k
LL_k ≡ leasing of land in period k

Constraints:

1 ≡ land in October–February (period 1)
2 ≡ land in March–May (period 2)
3 ≡ land in June–September (period 3)
4 ≡ maximum land rentable in period 1
5 ≡ maximum land rentable in period 2
6 ≡ maximum land rentable in period 3

The peculiarity of this problem is that it simulates a sequence of time periods within which activities must be allocated. To facilitate the task of setting up the LP model correctly, a simple diagram may be convenient.

The equality sign in the constraints for the land owned by the farmer is justified by his desire to lease all the unused land in each period. The revenue coefficients in the initial tableau are measured in thousand of dollars.

Primal

maximize $NR = 500W + 1500T + 1300CT + 2000CL + 200LL_1 + 200LL_2$
$\qquad +200LL_3 - 200RL_1 - 200RL_2 - 200RL_3$

subject to

$$
\begin{array}{llll}
W & + CL + LL_1 & - RL_1 & = & 500 \\
W & +T & + LL_2 & - RL_1 - RL_2 & = & 500 \\
& T + CT & + LL_3 - RL_1 - RL_2 - RL_3 = & 500 \\
& & RL_1 & \leq & 100 \\
& & RL_2 & \leq & 100 \\
& & RL_3 \leq & 100
\end{array}
$$

$$\text{all variables} \geq 0$$

Initial Tableau

NR	W	T	CT	CL	LL_1	LL_2	LL_3	RL_1	RL_2	RL_3	sol
	1			1	1			-1			$= 500$
	1	1				1		-1	-1		$= 500$
		1	1				1	-1	-1	-1	$= 500$
								1			≤ 100
									1		≤ 100
										1	≤ 100
1	$-.5$	-1.5	-1.3	-2.0	$-.2$	$-.2$	$-.2$	$.2$	$.2$	$.2$	$= 0$

The economic interpretation follows the scheme outlined in previous examples and is left to the reader.

The optimal production plan for this farm includes 600 acres of cauliflower, 700 acres of tomatoes, and 100 acres of cantaloupe. The plan calls for renting 100 acres of land in each period. The maximum revenue to count on is $2,320,000.

The shadow prices for the initially available land are $2,000 per acre in period 1, $200 in period 2, and $1,300 in period 3. The value marginal products of rented land are $3,300 per acre in period 1, $300 in period 2, and $1,100 in period 3. The land shadow prices indicate that the availability of land in period 1 is most crucial, since the corresponding shadow price is by far the highest among the three periods.

Dual minimize $TC = 500y_1 + 500y_2 + 500y_3 + 100y_4$
$+ 100y_5 + 100y_6$

subject to

$$
\begin{aligned}
y_1 + y_2 & & & & & \geq 500 \\
y_2 + y_3 & & & & & \geq 1500 \\
y_3 & & & & & \geq 1300 \\
y_1 & & & & & \geq 2000 \\
y_1 & & & & & \geq 200 \\
y_2 & & & & & \geq 200 \\
y_3 & & & & & \geq 200 \\
-y_1 - y_2 - y_3 + y_4 & & & & \geq -200 \\
- y_2 - y_3 + y_5 & & & \geq -200 \\
- y_3 + y_6 & & \geq -200
\end{aligned}
$$

y_1, y_2, and y_3 are free; y_4, y_5, and $y_6 \geq 0$

This example illustrates the meaning of relative tightness among primal constraints and the essential information that can be obtained from the dual problem: although all land constraints are binding or, in other words, all the available land is allocated to productive activities in each period, the degree of binding is different among them. The corresponding dual variables can then be interpreted as indices of relative tightness, a kind of information that cannot be inferred from the primal problem.

3.8 Problem 7: A Detailed Example

Previous examples have considered stylized specifications of production plans for entire farms. By necessity, the formulation of the problems had to be limited to a few technological aspects. In the present example, on the contrary, we would like to present the specification of the technology for one activity in far greater detail. We deal, therefore, with a partial problem to illustrate that a linear programming model can be assembled in blocks as detailed as desired. The following description thus is not a complete speci-fication of a linear programming problem but a more realistic formulation of technology for one activity.

Description. In a farm production plan the tomato activity requires (among other things) the following operations: cultivation, irrigation, and spraying, in that order. There are three periods of time (say weeks) for executing these operations. Cultivation, however, must be performed in period 1 (P1); irrigation may be performed in period 1 if time is available after cultivation and must be completed within the second period; spraying may be done in both period 1 and 2 if time is available after cultivation and irrigation and must be completed before expiration of period 3 (P3).

Cultivation requires 4 hours per unit of activity. Irrigation requires 2 hours if performed in period 1 and 3 hours if performed in period 2. Spraying requires .5, .75, and 1 hour if performed in period 1, 2, and 3, respectively. The number of hours available in the three periods is T_1, T_2, and T_3, respectively.

The task is to set up the correct specification of the activities and constraints.

Solution

Activities:

$x_1 \equiv$ tomatoes harvesting measured in acres
$x_2 \equiv$ cultivation in period 1 measured in acres
$x_3 \equiv$ irrigation in period 1 measured in acres
$x_4 \equiv$ irrigation in period 2 measured in acres
$x_5 \equiv$ spraying in period 1 measured in acres
$x_6 \equiv$ spraying in period 2 measured in acres
$x_7 \equiv$ spraying in period 3 measured in acres

Constraints:

$1 \equiv$ spraying
$2 \equiv$ possibility of irrigating in P1 and P2
$3 \equiv$ possibility of spraying in P1, P2, and P3
$4 \equiv$ possibility of spraying in P1
$5 \equiv$ time available in P1
$6 \equiv$ time available in P2
$7 \equiv$ time available in P3

First of all, if tomatoes are profitable, they must be cultivated, irrigated, and sprayed. Obviously, the harvested acreage cannot be greater than either the cultivated, the irrigated, or the sprayed acreage. It can, however, be less than those acreages, and the harvesting activity is contingent upon the spraying activity being performed. Hence, constraint 1 is specified as

harvested tomatoes $\leq$ sprayed tomatoes in P1, P2, and P3
$$x_1 \leq x_5 + x_6 + x_7$$

Secondly, the technology requires that operations be performed in sequence. Thus, irrigation comes after cultivation, and there is no point in irrigating more acreage than is cultivated. Therefore,

irrigated acreage in P1 and P2 $\leq$ cultivated acreage
$$x_3 + x_4 \leq x_2$$

Similarly, spraying must come after cultivation and irrigation. Because of the previous constraint, if it comes after irrigation it will come also after cultivation. Hence,

sprayed acreage in P1, P2, and P3 $\leq$ irrigated acreage in P1 and P2
$$x_5 + x_6 + x_7 \leq x_3 + x_4$$

Because of the nested possibilities for performing operations in the various periods, we must make sure that they are not executed out of order. That is, spraying in P1 cannot come before irrigation in P2, something that the previous constraint would admit. Thus, another constraint involving sprayed and irrigated acreage is required as

sprayed acreage in P1 $\leq$ irrigated acreage in P1
$$x_5 \leq x_3$$

Finally, the operations of cultivation, irrigation, and spraying require that the total time necessary for their execution be less than or equal to the time available in each period:

$$
\begin{aligned}
4x_2 + 2x_3 \quad + .5x_5 \quad &\leq T_1 \\
3x_4 + \quad .75x_6 \quad &\leq T_2 \\
x_7 &\leq T_3
\end{aligned}
$$

The complexity of the technology description for the tomato activity is displayed in the following initial tableau.

Initial Tableau

x_1	x_2	x_3	x_4	x_5	x_6	x_7		sol
1				−1	−1	−1	$\leq$	0
	−1	1	1				$\leq$	0
		−1	−1	1	1	1	$\leq$	0
		−1		1			$\leq$	0
	4	2		.5			$\leq$	T_1
			3		.75		$\leq$	T_2
						1	$\leq$	T_3

Observe that without a rational guideline for specifying the appropriate constraints it would have been rather difficult, or altogether impossible, to formulate directly an initial tableau such as the one above. In closing this chapter, therefore, we emphasize the initial recommendation for formulating correct linear programming models: first, identify activities and constraints; second, relate activities to constraints using the general scheme

$$\boxed{\text{consumption} + \text{investment} \leq \text{production} + \text{availability}}$$

where all terms are associated with a + sign; third, rearrange the variables on one side of the constraints and, only at that stage, set up the initial tableau.

Chapter 4

The Transportation Problem

4.1 Introduction

To further motivate the structure and the economic interpretation of a linear programming model, we will now discuss a transportation problem. Together with the minimum-cost diet problem of Stigler, the transportation problem was among the first formulations of linear programming problems. The Russian L. V. Kantorovich formulated the first specification of the problem in 1939, but his work became known in the West about a decade later. In 1941, Frank L. Hitchcock wrote a paper in the *Journal of Mathematics and Physics* whose title seems more fit for a journal of economics: "The Distribution of a Product from Several Sources to Numerous Localities." In that paper he also gave the correct solution of a numerical example by means of an algorithm, which is probably the first solution method for a linear programming problem.

As Hitchcock's title indicates, a transportation problem exists whenever a flow of commodities must be transported over several possible routes connecting localities, called *origins* and *destinations,* and the cost of transporting the commodities is to be minimized.

Description. A trucking company must transport a given product from m origins to n destinations. Sometimes origins are also called sources and warehouses, while destinations are called markets. The job must be carried out at the minimum total cost.

Solution. The formulation of the problem will be given in general terms. Thus, let

$c_{ij} \equiv$ transport cost of one unit of product from origin i to destination j

$x_{ij} \equiv$ amount of product shipped from origin i to destination j

Then, the objective of the trucking firm's manager can be represented as

$$\text{minimize total cost of shipment } = \sum_{i=1}^{m} \sum_{j=1}^{n} c_{ij} x_{ij}$$

The contract specifies the following constraint for each destination j:

amount shipped to destination j from all origins $\geq$ amount ordered at destination j

In symbols, these constraints can be written as

$$\sum_{i=1}^{m} x_{ij} \geq b_j, \quad j = 1, \ldots, n$$

where b_j is the amount of product demanded at destination j. The quantity on the left-hand side of the inequality sign can also be interpreted as the supply of the product in question. Hence, the above restriction is a condition of material equilibrium reflecting the fundamental relation, demand $\leq$ supply.

Another obvious set of constraints refers to origins:

amount shipped from origin i to all destinations $\leq$ amount available at origin i

Again, in symbols:

$$\sum_{j=1}^{n} x_{ij} \leq a_i, \quad i = 1, \ldots, m$$

where a_i is the amount of product available at origin i. It is also required that all the amounts of product shipped over the various routes be nonnegative.

The above formulation is the primal specification of the transportation problem, which is reassembled below in a more compact form.

Primal minimize $TC = \sum_{i=1}^{m}\sum_{j=1}^{n} c_{ij}x_{ij}$

subject to $\sum_{i=1}^{m} x_{ij} \geq b_j, \qquad j = 1,\ldots,n$

$$-\sum_{j=1}^{n} x_{ij} \geq -a_i, \qquad i = 1,\ldots,m$$

$$x_{ij} \geq 0 \text{ for all } i\text{'s and } j\text{'s}$$

All the constraints were written in the same direction to facilitate the formulation of the dual problem. Before attacking this task, however, it is important to consider the initial tableau associated with the primal transportation problem, which possesses a special structure.

Initial Tableau

x_{11}	x_{12}	x_{1n}	x_{21}	x_{22}	x_{2n}	x_{m1}	x_{m2}	x_{mn}		
1	0...	0	1	0...	0...	1	0...	0	$\geq$	b_1
0	1...	0	0	1...	0...	0	1...	0	$\geq$	b_2
0	0...	1	0	0...	1...	0	0...	1	$\geq$	b_n
-1	-1...	-1	0	0...	0...	0	0...	0	$\geq$	$-a_1$
0	0...	0	-1	-1...	-1...	0	0...	0	$\geq$	$-a_2$
0	0...	0	0	0...	0...	-1	-1...	-1	$\geq$	$-a_m$

All the elements of the transportation matrix are either zero or unity. Each column of the initial tableau contains exactly two nonzero coefficients. This structure greatly facilitates the computations of the solution.

4.2   The Dual of the Transportation Problem

Although we already know how to write down algebraically the dual of LP problems, let us for a moment put aside this knowledge and think entirely in economic terms. It seems natural, in this case, to specify the same transportation problem in a different form or, better, from a different viewpoint. Suppose, in fact, that the shipping contract calls for the trucking company to actually buy the product at the origins and sell it at the destinations at whatever prices it is able to pay and receive, given that the transportation costs on all routes are fixed and known. The problem for the manager of the trucking company, then, will be: what prices p_j^d and p_i^o should the manager be prepared to take and offer at the destinations and origins, respectively? A reasonable answer could be the following one: those prices that maximize the difference between the value of the goods sold at the destinations and the value of the same goods bought at the origins. This answer could be interpreted as the maximization of the value added of the trucking firm or the transportation industry. This second objective, therefore, can be formalized as follows:

$$\text{maximize} \quad \begin{bmatrix} \text{value of goods} & & \text{value of goods} \\ \text{transported and sold} & - & \text{bought} \\ \text{at the destinations} & & \text{at the origins} \end{bmatrix}$$

or, in symbols:

$$\text{maximize} \quad \left[\sum_{j=1}^{n} b_j p_j^d \ - \ \sum_{i=1}^{m} a_i p_i^o \right]$$

The reassembled specification of the problem will show clearly that we have described the dual of the transportation problem:

Dual maximize $VA = \left[\displaystyle\sum_{j=1}^{n} b_j p_j^d \ - \ \sum_{i=1}^{m} a_i p_i^o \right]$

$$\text{subject to} \qquad p_j^d - p_i^o \leq c_{ij} \text{ for all } i\text{'s and } j\text{'s}$$

$$p_j^d \text{ and } p_i^o \geq 0$$

The primal and dual problems described above lead in the same direction and represent two equivalent ways to reach the same economic welfare of the trucking company. Notice that each column of the initial tableau corresponds to a row in the dual constraints, as required by the mathematical structure of linear programming.

4.3   A Numerical Transportation Problem

Description. A trucking company has contracted to supply weekly three Lucky supermarkets located in Modesto, Sonoma, and Fresno, California, with potatoes kept at two Lucky warehouses in Sacramento and Oakland, also in California. The weekly availability (supply) of potatoes at the warehouses is 400 and 600 tons, respectively. The supermarkets' weekly needs

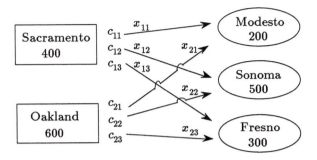

Figure 4.1. The flow of information in a transportation problem.

(demands) are 200, 500, and 300 tons of potatoes, respectively. The transportation unit costs are given in table 4.1.

The task is to find the supply routes that minimize the total transportation costs for the trucking firm. The contract stipulates that all supermarkets must receive at least the quantity of potatoes specified above.

Solution. The description of the problem indicates that there are two origins (warehouses), Sacramento and Oakland, and three destinations (markets), Modesto, Sonoma, and Fresno. For the purpose of recognizing the flow of information, it is convenient to represent the linking routes among warehouses and markets as in figure 4.1.

The formulation of the primal problem requires the specification of two sets of constraints: the demand constraints and the availability constraints. The demand constraints stipulate that the demand of potatoes by each market be satisfied by the supply from all warehouses:

$$x_{11} + x_{21} \geq 200 \equiv b_1 \quad \text{Modesto market}$$
$$x_{12} + x_{22} \geq 500 \equiv b_2 \quad \text{Sonoma market}$$
$$x_{13} + x_{23} \geq 300 \equiv b_3 \quad \text{Fresno market}$$

The availability constraints require that each warehouse not ship out more

Table 4.1. Transportation Unit Costs (dollars per ton)

Warehouse	Supermarkets		
	Modesto	Sonoma	Fresno
Sacramento	40	50	80
Oakland	40	30	90

of the commodity than is available on site:

$$x_{11} + x_{12} + x_{13} \leq 400 \equiv a_1 \qquad \text{Sacramento warehouse}$$
$$x_{21} + x_{22} + x_{23} \leq 600 \equiv a_2 \qquad \text{Oakland warehouse}$$

The objective is to minimize the total transportation cost:

$$\text{minimize } TC = 40x_{11} + 50x_{12} + 80x_{13} + 40x_{21} + 30x_{22} + 90x_{23}$$

Each variable enters each set of constraints only once. Therefore, the transportation matrix describing the technology of transportation is represented in the following initial tableau where the warehouse constraints have been reversed in order to maintain the same direction of the inequalities.

Initial Tableau

TC	x_{11}	x_{12}	x_{13}	x_{21}	x_{22}	x_{23}		sol
	1			1			$\geq$	200
		1			1		$\geq$	500
			1			1	$\geq$	300
	-1	-1	-1				$\geq$	-400
				-1	-1	-1	$\geq$	-600
1	-40	-50	-80	-40	-30	-90	$=$	0

From the above initial tableau it is easy to state the dual problem:

Dual

$$\text{maximize } VA = 200p_1^d + 500p_2^d + 300p_3^d - 400p_1^o - 600p_2^o$$

$$p_1^d \qquad\qquad\qquad - \quad p_1^o \qquad\qquad \leq 40$$
$$p_2^d \qquad\qquad - \quad p_1^o \qquad\qquad \leq 50$$
$$p_3^d \quad - \quad p_1^o \qquad\qquad \leq 80$$
$$p_1^d \qquad\qquad\qquad - \quad p_2^o \leq 40$$
$$p_2^d \qquad\qquad - \quad p_2^o \leq 30$$
$$p_3^d \qquad - \quad p_2^o \leq 90$$

$$p_j^d \text{ and } p_i^o \geq 0 \text{ for all } i\text{'s and } j\text{'s}$$

The economic interpretation of this dual example is similar to that given for the general specification.

Chapter 5

Spaces and Bases

5.1 Introduction

In previous chapters we learned how to set up LP problems and the proper relations between their primal and dual specifications. In this chapter we will examine more closely the geometric and algebraic structure of the linear programming model. For ease of exposition and graphing, we will discuss a very simplified linear programming problem with only two outputs and two inputs. Every notion and result presented here, however, can be extended to more complex LP formulations with many hundreds and even thousands of outputs and inputs.

Consider the following primal LP problem:

Primal maximize $Z = 4x_1 + 5x_2$

subject to

$$2x_1 + 6x_2 \leq 12 \leftarrow \text{land (input)}$$
$$4x_1 + x_2 \leq 8 \leftarrow \text{labor (input)}$$

$$\begin{array}{cc} \uparrow & \uparrow \\ \text{wheat} & \text{corn} \\ \text{activity} & \text{activity} \\ \text{(output)} & \text{(output)} \end{array}$$

$$x_1 \geq 0, \text{ and } x_2 \geq 0$$

The nonnegativity constraints imposed upon the primal variables x_1 and x_2 are specified apart from the main structural constraint as a matter of convention established over the past forty years. The idea is that these constraints are so simple and universal that it is possible to write down a linear programming problem without taking too much space in order to list them. They are, nevertheless, important constraints which cannot

67

be forgotten under any circumstance. Furthermore, this problem admits slack variables corresponding to the unused amount of each limiting input, although they are not explicitly indicated. To ensure that the reader is becoming conversant with establishing an immediate relation with a LP problem admitting slack variables, we will restate the same problem.

Primal

maximize $Z = \quad 4x_1 \quad + \quad 5x_2$

subject to
$$2x_1 \quad + \quad 6x_2 \quad + \quad x_{s1} \qquad\qquad = 12 \leftarrow\text{land (input)}$$
$$4x_1 \quad + \quad x_2 \qquad\qquad + \quad x_{s2} \quad = \quad 8 \leftarrow\text{labor (input)}$$

↑	↑	↑	↑
wheat	corn	land	labor
activity	activity	slack	slack
(output)	(output)	activity	activity

$$x_1 \geq 0,\ x_2 \geq 0,\ x_{s1} \geq 0,\ \text{and}\ x_{s2} \geq 0$$

Slack variables do not appear in the objective function because their coefficient is zero. In economic terms, this condition reflects the notion of a free good: unused land and labor have zero value.

5.2 Two Ways of Looking at a System of Constraints

The system of constraints in the above problem can be graphed in two different but related ways. The first graph has the variables x_1 and x_2 labeling the coordinate axes. Because these variables represent the levels of corn and wheat *outputs,* the constraints of the problem will be graphed in the *output space,* as in figure 5.1. The second graph (figure 5.2) shows the coordinate axes labeled by the *inputs* and, therefore, will be called the *input requirement space.*

The graphing of the primal problem in the output space is easily performed by graphing each constraint (including the nonnegativity constraints) in the first formulation of the LP problem considered as an equality. It is important to represent the direction of the inequality in each primal constraint. Such a direction is given by a vector perpendicular (called also orthogonal or normal) to the line defined by the constraint. The intersection of all primal constraints (the intersection of all half spaces, in more technical terms) forms a *feasible region,* called also *admissible region.* In a linear programming problem the feasible region is always a *polygon,* that is, a geometrical object with a finite number of *faces, edges,* and *extreme points* (corner points). The extreme points are especially important because

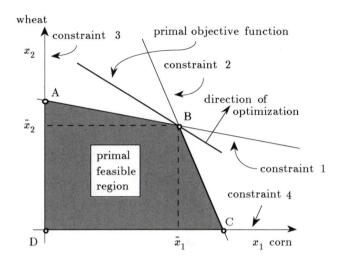

Figure 5.1. The output space.

they form the basis for the solution of a linear programming problem, as explained in chapter 6.

In looking at the graph of the output space, a legitimate question can be formulated: can we say anything about the slack (surplus) variables? In other words, can we read the level of the slack variables off figure 5.1? The answer is affirmative: we can read the levels of the slack variables *in qualitative terms.*

Consider, for example, point B, which lies on both constraints #1 and #2. Relative to point B, $x_{s1} = x_{s2} = 0$; that is, both slack variables have zero value. Consider point C, which lies on constraint #2 but not on constraint #1. In this case, $x_{s1} > 0$ and $x_{s2} = 0$. We can only say that the slack variable x_{s1} is strictly greater than zero, thus giving a qualitative representation of its level. In some sense, the slack variable x_{s1} measures the distance of the point C from constraint #1. Finally, relative to point D, the slack variables are $x_{s1} > 0$ and $x_{s2} > 0$.

The graph of the primal problem into the input requirement space is based upon a further reformulation of the system of primal constraints centered upon the notion of *vectors* as follows:

$$\begin{bmatrix} 2 \\ 4 \end{bmatrix} x_1 + \begin{bmatrix} 6 \\ 1 \end{bmatrix} x_2 + \begin{bmatrix} 1 \\ 0 \end{bmatrix} x_{s1} + \begin{bmatrix} 0 \\ 1 \end{bmatrix} x_{s2} = \begin{bmatrix} 12 \\ 8 \end{bmatrix}$$
$$\quad \uparrow \qquad\qquad \uparrow \qquad\qquad \uparrow \qquad\qquad \uparrow \qquad\qquad \uparrow$$
$$\quad \mathbf{a_1} \qquad\quad \mathbf{a_2} \qquad\quad \mathbf{e_1} \qquad\quad \mathbf{e_2} \qquad\quad \mathbf{b}$$

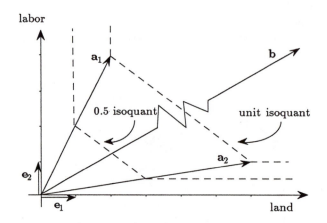

Figure 5.2. The input requirement space.

The coefficients associated with each variable have been isolated into boxes called vectors. In figure 5.2, the five vectors a_1, a_2, e_1, e_2, and b are graphed using the corresponding elements as coordinates in the input space. In the rest of this book a bold character will indicate a vector.

The input requirement space contains the map of isoquants corresponding to the linear technology specified by the LP problem. In traditional economic terminology, an isoquant is the locus of input combinations that can produce the same (iso) level(s) [quant(s)] of output(s). In figure 5.2, the dashed and kinked line connecting the tips of the arrows associated with a_1 and a_2 represents the unit isoquant because it is the locus of input points for which $x_1 + x_2 = 1$. The solution of the LP problem in the input requirement space must involve the vector of input availabilities, b. Hence, to better understand the structure and the meaning of a linear programming problem graphed in the input requirement space we need a digression on vectors and bases.

5.3   A Primer on Vectors and Bases

A *vector* is a rectangular box containing the coefficients of an activity in an orderly fashion: the coefficient of the first constraint in the first position and the coefficient of the second constraint in second position. We say that

$$a_1 = \begin{bmatrix} 2 \\ 4 \end{bmatrix}$$

is a vector in E^2 (the Euclidian space of 2 dimensions) and we graph it as in figure 5.2. The coordinates of the point coincident with the tip of the arrow associated with a_1 are 2 and 4, respectively. All other vectors in figure 5.2 have similar structure and graphical interpretation. Special types of vectors are

$$e_1 = \begin{bmatrix} 1 \\ 0 \end{bmatrix}$$

and

$$e_2 = \begin{bmatrix} 0 \\ 1 \end{bmatrix}$$

associated with slack variables. These vectors are called *unit vectors*, and the subscript refers to the position of the unit element in the vector.

For the development that follows, three important operations can be performed with vectors:

1. Addition of two vectors.
2. Subtraction of two vectors.
3. Multiplication of a vector by a scalar.

5.3.1 Addition of Two Vectors

As can be gleaned from figure 5.3, the operation of adding two vectors can be given the geometric interpretation of constructing a parallelogram whose main diagonal is the vector resulting from the addition.

Consider the following vector addition as an example:

$$\begin{bmatrix} 3 \\ -1 \end{bmatrix} + \begin{bmatrix} 2 \\ 4 \end{bmatrix} = \begin{bmatrix} 5 \\ 3 \end{bmatrix}$$

$$\uparrow \qquad \uparrow \qquad \uparrow$$
$$\mathbf{v_1} \quad + \quad \mathbf{v_2} \quad = \quad \mathbf{v_3}$$

In figure 5.3, this addition corresponds to the main diagonal, $\mathbf{v_3}$, of the parallelogram, whose sides are the vectors $\mathbf{v_1}$ and $\mathbf{v_2}$. Notice the important point: the elements of $\mathbf{v_3}$ are the horizontal summations of the corresponding elements of $\mathbf{v_1}$ and $\mathbf{v_2}$.

5.3.2 Subtraction of Two Vectors

Subtraction of two vectors is similar to their addition, since

$$\mathbf{v_1} - \mathbf{v_2} = \mathbf{v_1} + (-\mathbf{v_2}) = \mathbf{v_3}$$

This operation is illustrated in figure 5.4. It requires thinking of the vector $(-\mathbf{v_2})$ as the vector $\mathbf{v_2}$ multiplied by (-1).

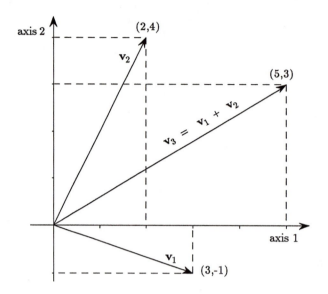

Figure 5.3. Vector addition.

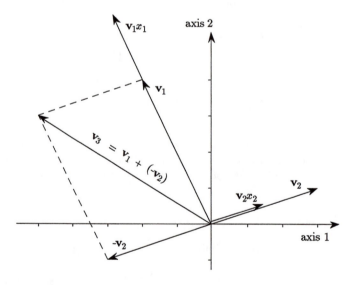

Figure 5.4. Subtraction of two vectors and multiplication by scalar.

5.3.3 Multiplication of a Vector by a Scalar

The operation of multiplying a vector $\mathbf{v}_1$ by a scalar x_1 can be interpreted as "stretching" or "shrinking" the given vector $\mathbf{v}_1$, depending on whether the scalar x_1 is greater or less than unity. Figure 5.4 illustrates this operation for the following two vectors:

$$\mathbf{v}_1 x_1 \text{ for } x_1 > 1, \text{ where } \mathbf{v}_1 = \begin{bmatrix} -2 \\ 4 \end{bmatrix} \text{ and } x_1 = \frac{3}{2}$$

$$\mathbf{v}_2 x_2 \text{ for } 0 < x_2 = \frac{1}{2} < 1 \text{ and } x_2 = -1, \text{ where } \mathbf{v}_2 = \begin{bmatrix} 3 \\ 1 \end{bmatrix}$$

The operation of multiplying vector $\mathbf{v}_2$ by $x_2 = -1$ results in the same vector oriented in the opposite direction.

5.3.4 Extreme Points

Consider the output space of figure 5.1. Points A, B, C, and D are technically called *extreme points* of the set of feasible solutions. Often, they are also called vertices. The notion of extreme point is crucial in linear programming because the optimal solution always occurs at an extreme point, as illustrated in section 1.9.

5.3.5 Basis

We are finally ready to introduce the very important notion of *basis*. Its importance is derived from the fact that in order to operate within a space a basis is required. We can give several definitions of a basis, some more formal than others, but all capable of illustrating the essential role played by a basis in everyday life and also in linear programming. Hence, a basis is

    1.   A system of measurement units.
    2.   A frame of reference.
    3.   A scaffolding of space.
    4.   A set of linearly independent vectors that span (or generate) the space.

Definition 1: A system of measurement units. The metric system forms a basis useful for measuring objects in everyday life. The same objects can be measured also by the standard American system, which constitutes another basis for the same universe (space) of objects. It is possible to change from the metric to the standard American system and vice versa. This operation represents a change of basis.

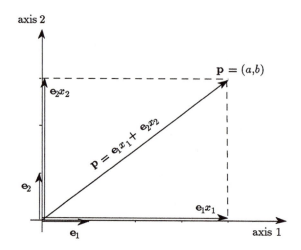

Figure 5.5. The point $\mathbf{p} = (a, b)$ expressed with the Cartesian basis.

Definition 2: A frame of reference. This definition can be illustrated by figure 5.5. Suppose we are given a point $\mathbf{p} = (a, b)$. To establish its location in a communicable way, we must introduce an accepted coordinate system such as the Cartesian system. Hence, the Cartesian system of coordinates is nothing else than a reference basis providing a system of measurement affine to the metric system. In a two-dimensional space, the Cartesian basis can be represented as

$$\text{Cartesian basis} = \left\{ \begin{bmatrix} 1 \\ 0 \end{bmatrix} \begin{bmatrix} 0 \\ 1 \end{bmatrix} \right\}$$

$$\uparrow \qquad \uparrow$$
$$\mathbf{e}_1 \qquad \mathbf{e}_2$$

Conventionally, the $\mathbf{e}_1$ vector represents the unit arrow on the horizontal axis and, similarly, $\mathbf{e}_2$ is the unit vector on the vertical axis. Then, the point $\mathbf{p} = (a, b)$ can be measured with reference to the Cartesian basis $(\mathbf{e}_1, \mathbf{e}_2)$ as

$$\begin{bmatrix} 1 \\ 0 \end{bmatrix} x_1 + \begin{bmatrix} 0 \\ 1 \end{bmatrix} x_2 = \begin{bmatrix} a \\ b \end{bmatrix}$$
$$\mathbf{e}_1 x_1 \quad + \quad \mathbf{e}_2 x_2 \quad = \quad \mathbf{p}$$

Expressed in this way, the point $\mathbf{p}$ can be interpreted as a vector that is the main diagonal of the parallelogram with sides $\mathbf{e}_1 x_1$ and $\mathbf{e}_2 x_2$. Therefore, from the above Cartesian expression of the point **p**, it is easy to verify that $x_1 = a$ and $x_2 = b$. We also say that the point **p** is expressed in terms of the basic vectors (e_1, e_2). All this discussion is illustrated in figure 5.5.

We wish now to measure the same point **p** with reference to a new basis such as

$$\text{new basis} = \left\{ \begin{bmatrix} 2 \\ -3 \end{bmatrix} \quad \begin{bmatrix} 3 \\ 5 \end{bmatrix} \right\}$$
$$\qquad\qquad\qquad \uparrow \qquad \uparrow$$
$$\qquad\qquad\qquad \mathbf{a}_1 \qquad \mathbf{a}_2$$

The vector **p**, in other words, must be expressed in terms of the new basis vectors ($\mathbf{a}_1$, $\mathbf{a}_2$), as in the following system of equations:

$$\mathbf{a}_1 x_1 \quad + \quad \mathbf{a}_2 x_2 \quad = \quad \mathbf{p}$$
$$\begin{bmatrix} 2 \\ -3 \end{bmatrix} x_1 + \begin{bmatrix} 3 \\ 5 \end{bmatrix} x_2 = \begin{bmatrix} a \\ b \end{bmatrix}$$

Again, the vector **p** is the main diagonal of the parallelogram constructed using ($\mathbf{a}_1$, $\mathbf{a}_2$), as illustrated in figure 5.6. It should be clear, therefore, that the same point (vector) can be expressed with reference to different bases. The "stretching" and "shrinking" coefficients (x_1, x_2) associated with the two bases have different values and constitute the solution of the measurement problem. Finally, a basis always forms an array of numbers that constitutes a square matrix.

Definition 3: A scaffolding of space. This definition is inherent in the discussion outlined above. As figures 5.5 and 5.6 illustrate, without the Cartesian and the other bases there could not be any way to measure point **p**. Hence, the basis must be in place prior to the measurement (solution) as the scaffolding must be in place prior to the erection of a building.

Definition 4: A set of linearly independent vectors that span the space. This last definition of basis is a formal one. We want to keep this definition in mind and use it throughout the remainder of the book. The other three definitions were presented only as a way to clarify the meaning and the role of a basis.

For a two-dimensional space E^2, a basis is a set of two linearly independent vectors that span the space. We also say that the same vectors generate the space because they can express (generate) any point in the E^2 space as a linear combination, as outlined in the discussion of definition 2. Similarly, for an m-dimensional space, a basis is a set of m linearly independent vectors that span the space.

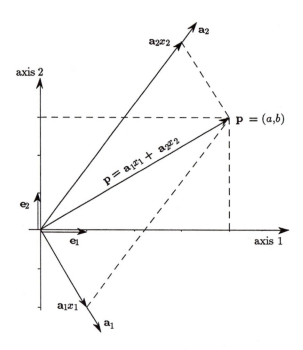

Figure 5.6. The point $\mathbf{p} = (a, b)$ expressed with a new basis.

But what is *linear independence* and how can it be verified? Linear independence is a mathematical notion involving the position of vectors relative to each other. It can be verified graphically and algebraically.

Graphically: Consider the input requirement space of figure 5.2. Two vectors that are linearly independent should form an angle different from $0°$ and $180°$. In other words, two linearly independent vectors cannot be parallel. Therefore, $(\mathbf{a}_1, \mathbf{a}_2)$, $(\mathbf{a}_2, \mathbf{e}_1)$, $(\mathbf{e}_1, \mathbf{e}_2)$, etc., are sets of linearly independent vectors that span the space and, as a consequence, they are bases. Obviously, the graphical verification of linear independence can be done for spaces of very low dimension. For spaces of dimension higher than two or three, the verification of linear independence must be done using other means.

Algebraically: In order to check whether two vectors, say $(\mathbf{a}_1, \mathbf{a}_2)$, form a basis for E^2, construct the array (matrix) of coordinates of the two vectors, that is,

$$[\mathbf{a}_1, \mathbf{a}_2] = \begin{bmatrix} 2 & 6 \\ 4 & 1 \end{bmatrix}$$

and compute the *determinant* (det) of this matrix. If the determinant of the matrix equals zero, the vectors are linearly dependent. If det $\neq 0$, the vectors are linearly independent.

The determinant of a (square) matrix is defined as follows:

$$\det \begin{bmatrix} a_{11} & a_{12} \\ a_{21} & a_{22} \end{bmatrix} = a_{11}a_{22} - a_{12}a_{21}$$

$$\det \begin{bmatrix} 2 & 6 \\ 4 & 1 \end{bmatrix} = 2 \times 1 - 6 \times 4 = -22 \neq 0$$

Hence, the above matrix constitutes a basis for E^2. For a correct calculation of a determinant, notice the order and direction of multiplication of the matrix elements.

Back to figure 5.2. How many bases (excluding the **b** vector) can we find in this problem? This many:

$$[a_1, a_2] \quad [a_1, e_1] \quad [a_1, e_2] \quad [a_2, e_1] \quad [a_2, e_2] \quad [e_1, e_2]$$

$$\begin{bmatrix} 2 & 6 \\ 4 & 1 \end{bmatrix} \begin{bmatrix} 2 & 1 \\ 4 & 0 \end{bmatrix} \begin{bmatrix} 2 & 0 \\ 4 & 1 \end{bmatrix} \begin{bmatrix} 6 & 1 \\ 1 & 0 \end{bmatrix} \begin{bmatrix} 6 & 0 \\ 1 & 1 \end{bmatrix} \begin{bmatrix} 1 & 0 \\ 0 & 1 \end{bmatrix}$$

Verify algebraically that each pair of vectors forms a basis in E^2.

5.4 Feasible Bases and Feasible Solutions

Linear programming's interest in bases stems from the fact that the optimal solutions occur at extreme points (see the discussion in section 1.9), which, in turn, are associated with bases. Indeed, the extreme points of interest are associated with a special kind of bases called *feasible bases*. It is important to emphasize that all this discussion about extreme points and bases is directly related to the output space and to the input requirement space as presented in figures 5.1 and 5.2, respectively. Figure 5.1 exhibits extreme points while figure 5.2 shows bases. The question is, therefore, how to associate extreme points with their corresponding bases.

To answer this question, we must be able to establish the value of the coordinates for the solutions of the linear programming problem occurring at extreme points A, B, C, and D in figure 5.1. These solutions will be established in qualitative terms, as discussed in section 5.2. In fact, if we were able to establish directly a linear programming solution in quantitative terms, there would not be the need to talk about bases.

Let us consider extreme point B, for example. It is easy to see that, at that point, $x_1 > 0$, $x_2 > 0$, $x_{s1} = x_{s2} = 0$. The slack variables are equal to zero because (x_1, x_2) at B is a point that lies on both lines defining constraints #1 and #2. No positive slack is, therefore, admissible. The above discussion is an example of a linear programming solution in qualitative terms. Similar calculations can be performed for all the extreme points, as follows:

extreme point A: $\{x_1 = 0, \ x_2 > 0, \ x_{s1} = 0, \ x_{s2} > 0\}$
extreme point B: $\{x_1 > 0, \ x_2 > 0, \ x_{s1} = 0, \ x_{s2} = 0\}$
extreme point C: $\{x_1 > 0, \ x_2 = 0, \ x_{s1} > 0, \ x_{s2} = 0\}$
extreme point D: $\{x_1 = 0, \ x_2 = 0, \ x_{s1} > 0, \ x_{s2} > 0\}$

For each of these extreme point solutions only two components are greater than zero, while the other two are equal to zero. We conclude, therefore, that every extreme point solution is associated with a particular representation of the set of LP constraints in equality form. Consider, in fact, the set of LP constraints (which include slack variables) for the numerical example of section 5.1. Using the qualitative information developed for each extreme point, we can associate every extreme point with a basic system of equations, as illustrated in the chart to the right.

We have achieved a remarkable result. The right-hand-side vector of constraint values has been expressed in terms of four different linear combinations of *basic vectors*. In other words, each system of equations corresponding to an extreme point utilizes a pair of vectors that form a basis. This is the reason for calling those systems of equations *basic systems*. Secondly, notice that adjacent extreme points correspond to bases that differ by only one vector. We conclude that to each extreme point in the output space (figure 5.1) there corresponds a pair of linearly independent vectors that form a basis in the input requirement space (figure 5.2).

When discussing the bases associated with this numerical example in section 5.4, we listed six of them, four of which appear in the above basic systems. Hence, among all the possible bases of a linear programming system of constraints, only those associated with extreme points of the set of feasible solutions are relevant for finding the optimal solution. These bases are called *feasible bases* because they return a *basic feasible solution*. Feasible bases, therefore, are associated with extreme points. In other words, feasible bases form the basic system of equations associated with extreme points, and the solution of these equations will provide a basic feasible solution of the LP problem in quantitative terms.

A further important conclusion can be derived from the above discussion. From a geometric viewpoint, a feasible basis in the input requirement space (figure 5.2) generates a cone that contains the constraint vector **b**. Bases such as $[\mathbf{a}_1, \ \mathbf{e}_2]$ and $[\mathbf{a}_2, \ \mathbf{e}_1]$ form cones that do not contain the **b** vector. The same bases do not appear in the basic systems of equations listed here. They do not correspond to any extreme point of the LP solution set and, therefore, they are not feasible bases.

Extreme Point $\longleftrightarrow$ **Basic System of Equations**

$$A \qquad \longleftrightarrow \qquad \begin{bmatrix} 6 \\ 1 \end{bmatrix} x_2 + \begin{bmatrix} 0 \\ 1 \end{bmatrix} x_{s2} = \begin{bmatrix} 12 \\ 8 \end{bmatrix}$$

$$6x_2 + 0x_{s2} = 12$$

or

$$x_2 + x_{s2} = 8$$

$$B \qquad \longleftrightarrow \qquad \begin{bmatrix} 2 \\ 4 \end{bmatrix} x_1 + \begin{bmatrix} 6 \\ 1 \end{bmatrix} x_2 = \begin{bmatrix} 12 \\ 8 \end{bmatrix}$$

$$2x_1 + 6x_2 = 12$$

or

$$4x_1 + x_2 = 8$$

$$C \qquad \longleftrightarrow \qquad \begin{bmatrix} 2 \\ 4 \end{bmatrix} x_1 + \begin{bmatrix} 1 \\ 0 \end{bmatrix} x_{s1} = \begin{bmatrix} 12 \\ 8 \end{bmatrix}$$

$$2x_1 + x_{s1} = 12$$

or

$$4x_1 + 0x_{s1} = 8$$

$$D \qquad \longleftrightarrow \qquad \begin{bmatrix} 1 \\ 0 \end{bmatrix} x_{s1} + \begin{bmatrix} 0 \\ 1 \end{bmatrix} x_{s2} = \begin{bmatrix} 12 \\ 8 \end{bmatrix}$$

$$x_{s1} + 0x_{s2} = 12$$

or

$$0x_{s1} + x_{s2} = 8$$

For maximum clarity we need to emphasize that a basic feasible solution is different from a feasible basis. They are two completely different (although related) entities. A basic feasible solution corresponds to an extreme point in the output space. A feasible basis corresponds to a set of linearly independent vectors (forming a cone around the constraint vector) in the input requirement space. To find the value of the basic feasible solution in numerical terms we must first know the appropriate feasible basis and then solve the corresponding basic system of equations. The fundamental process can be illustrated by the scheme in figure 5.7.

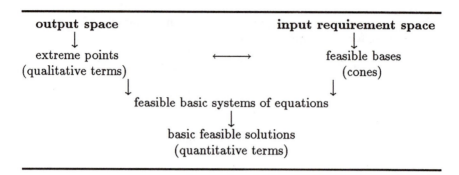

Figure 5.7. The flow of information in linear programming.

The entire discussion about feasible bases and basic feasible solutions has been developed with reference to the system of constraints of the primal problem using the output and the input requirement spaces. Analogously, the dual problem can be analyzed in terms of dual basic feasible solutions and dual feasible bases by means of two spaces: the marginal cost space and the marginal revenue space. The dual feasible bases have, in general, a different dimension from that of the primal feasible bases.

Generalization. Given the following linear programming problem,

Primal

$$\text{maximize } Z = \sum_{j=1}^{n} c_j x_j = c_1 x_1 + \ldots + c_n x_n$$

$$\text{subject to } \sum_{j=1}^{n} a_{1j} x_j = a_{11} x_1 + \ldots + a_{1n} x_n \leq b_1$$

$$\ldots\ldots\ldots\ldots\ldots\ldots\ldots\ldots\ldots\ldots\ldots\ldots\ldots\ldots\ldots$$

$$\sum_{j=1}^{n} a_{mj} x_j = a_{m1} x_1 + \ldots + a_{mn} x_n \leq b_m$$

$$x_j \geq 0, \quad j = 1, \ldots, n$$

we assume that the above system of constraints contains at least a basis for E^m. A basic feasible solution for the above problem, then, is a vector $\mathbf{x}' = [x_1, x_2, \ldots, x_n, x_{s1}, \ldots, x_{sm}]$ such that n components of this vector, not associated with basic vectors, are equal to zero and the remaining m components, associated with basic vectors, may be positive.

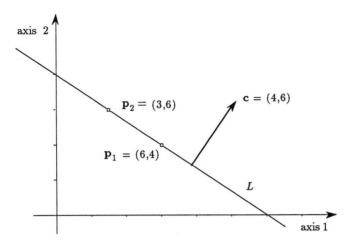

Figure 5.8. The dual definition of a line.

5.5 The Dual Definition of a Line or a Plane

We assume that the reader is familiar with the fact that two points uniquely define a line, three points a plane, four points a hyperplane, etc. This is the primal way of defining a line, a plane, etc. The dual way of defining these entities is as follows: a given vector c of coefficients defines the line, the plane, etc., that is perpendicular (orthogonal, normal) to it. In figure 5.8 the line L is defined in two equivalent ways: by the points p_1 and p_2, and by the vector c. Hence, given the line

$$4x_1 + 6x_2 = 48$$

the points $p_1 = (x_1 = 6,\ x_2 = 4)$, $p_2 = (x_1 = 3,\ x_2 = 6)$, are on line L. Also the vector $c = (4,\ 6)$ is the unique (up to a scalar) vector perpendicular to the line L. Since a vector represents a point in the parameter space, we conclude that the dual of a line is a point represented by a vector perpendicular to the given line. This notion is important because it reexplains the dual relationship between constraints and the associated variables, as presented in chapter 2. In fact, a constraint (a half space) is defined by a line, and a dual variable can be thought of as a point.

Chapter 6

Solving Systems of Equations

6.1 Introduction

The information about linear programming problems developed up to this point has revealed the need to know an efficient method for solving systems of equations. In fact, the main results achieved in previous chapters can be stylized as follows:

1. Optimal solutions occur always at an extreme point of the primal feasible region, as illustrated in the output space diagram of figure 5.1.

2. To each extreme point in the output space there corresponds a unique basis (for nondegenerate problems) in the input requirement space, as in figure 5.2.

3. To solve a linear programming problem, therefore, it is sufficient to explore only extreme points.

4. An extreme point of the primal solution set is associated with a basic system of equations whose solution provides a basic feasible solution in quantitative terms.

5. As a consequence, we will be discussing a solution procedure for linear programming problems called the simplex method, which is organized around a process of solving a *sequence* of systems of linear equations. A member of this sequence is called an *iteration*.

To summarize, because extreme points (one of which is optimal) are associated with basic feasible solutions and because these solutions reduce the LP inequalities to systems of basic linear equations, it is important to know an efficient procedure for solving systems of equations.

6.2 The Gaussian Elimination Method

Consider the following system of two equations in two unknowns:

$$(A) \qquad \begin{array}{rcl} 2x_1 + 4x_2 &=& 16 \\ 6x_1 + 2x_2 &=& 18 \end{array}$$

A solution for the above system of equations is a set of numbers $\bar{x}_1$ and $\bar{x}_2$ such that they satisfy each equation. The Gaussian (from Karl Friedrich Gauss, 1777–1855) elimination method transforms system (A) in such a way that, at the end, it will be easy to read off the solution. That is, the Gaussian algorithm transforms (A) into the following system:

$$(B) \qquad \begin{array}{rcl} x_1 &=& \bar{x}_1 \\ x_2 &=& \bar{x}_2 \end{array}$$

We notice that in system (B):

1. Variable x_2 was eliminated from the first equation.
2. Variable x_1 was eliminated from the second equation.
3. The coefficients of x_1 in the first equation and of x_2 in the second equation of system (A) have been normalized (have the value of 1) in (B).

To achieve the (B) form of the equation system, the following steps were taken:

Step 1: (a) To normalize the coefficient 2 of x_1 in the first equation of (A), multiply equation #1 by 1/2, the reciprocal of 2, to obtain:

$$\text{equation \#1:} \quad \begin{array}{rcl} 2x_1 + 4x_2 &=& 16 \\ (1/2)2x_1 + (1/2)4x_2 &=& (1/2)16 \\ x_1 + 2x_2 &=& 8 \end{array}$$

(b) To eliminate x_1 from equation #2 in (A), multiply equation #1 by -3 and add the result to equation #2:

multiplication: equation #1 $(-3)2x_1 + (-3)4x_2 = (-3)16$

$$\text{addition:} \quad \begin{array}{lrcl} \text{equation \#1} & -6x_1 - 12x_2 &=& -48 \\ \text{equation \#2} & \underline{6x_1 + 2x_2} &=& \underline{18} \\[4pt] \text{equation \#2} & -10x_2 &=& -30 \end{array}$$

$$\downarrow$$
$$x_1 \text{ is eliminated}$$

At this stage, system (A) has been transformed into the equivalent system:

$$(A^*) \qquad\qquad \begin{aligned} x_1 + 2x_2 &= 8 \\ -10x_2 &= -30 \end{aligned}$$

Step 2: The goal during this step is to eliminate x_2 from the first equation of (A^*) and to normalize the coefficient of x_2 in the second equation.

(a)   To normalize the coefficient (-10) of x_2 in equation #2 of (A^*), multiply the second equation by $(-1/10)$:

equation #2: $(-1/10)(-10)x_2 = (-1/10)(-30)$
$$x_2 = 3$$

(b)   To eliminate x_2 from equation #1 in (A^*), multiply equation #2 in (A^*) by 2/10 and add the result to equation #1:

multiplication: equation #2 $2/10(-10)x_2 = 2/10(-30)$

addition: equation #2 $-$ $2x_2 = -6$
               equation #1 $x_1 +$ $2x_2 =$ 8
               _____

            equation #1 x_1 $= 2$

$\downarrow$

x_2 is eliminated

From the results achieved in step 2 we have: $x_1 = 2$ [from (b)] and $x_2 = 3$ [from (a)]. This result is precisely the transformed system as desired in (B). The choice of coefficients for operating the transformation, however, was not a simple routine. Fortunately, there is a way to achieve order in this process. It is described by a procedure called the *pivot method* to be discussed next.

6.3   The Pivot Method

The presentation of the alternative procedure for solving a system of linear equations will use the same numerical example presented in the previous section. The method begins with the arrangement of the coefficients of system (A) in a tableau whose columns are labeled by the corresponding variables:

$$\begin{array}{ccc} x_1 & x_2 & \text{sol} \end{array}$$
$$\begin{bmatrix} 2 & 4 & 16 \\ 6 & 2 & 18 \end{bmatrix}$$

6.3.1   The Transformation Matrix

Next, we must define a *transformation matrix* T_1 whose purpose is to store all the information about the coefficients defined in step 1 of the previous section. The definition of T_1 is as follows: take a (2×2) *identity matrix,* that is

$$I = \begin{bmatrix} 1 & 0 \\ 0 & 1 \end{bmatrix}$$

and replace one of these two column vectors with a *transformation vector.* Which vector of the I matrix is to be replaced is dictated by the choice of the coefficient to be normalized, or the *pivot.* In step 1 of the previous section, we chose 2 of x_1 as the coefficient to be normalized and therefore the pivot is the same coefficient 2.

The following general rule applies to the definition of the transformation matrix:

> Place the transformation vector in that column of the I matrix whose index corresponds to the row containing the pivot.

Since we chose the pivot in the first equation (first row) of system (A), the transformation vector will replace the first column of I in the construction of the transformation matrix T_1.

The definition of the transformation vector follows this general rule:

> The reciprocal of the pivot is placed in the same row of the transformation vector as that of the pivot; the other coefficients in the pivot column are divided by the pivot and placed in the transformation vector with a changed sign.

Continuing the illustration of the numerical example used in the previous section we have the following setup corresponding to step 1:

$$\begin{array}{ccc} T_1 & \quad x_1 \quad x_2 \quad \text{sol} \\ \begin{bmatrix} 1/2 & 0 \\ -3 & 1 \end{bmatrix} & \begin{bmatrix} \boxed{2} & 4 & 16 \\ 6 & 2 & 18 \end{bmatrix} \end{array}$$

It is important to mark the pivot element in order to remember its position. In this example the pivot 2 has been boxed. Notice that the same coefficients $(1/2, -3)$ developed in step 1 of section 6.2 appear in the transformation (column) vector in T_1. This fact is consistent with the equivalence between the pivot method and the Gaussian elimination algorithm.

Computations are now carried out by multiplying each column of the second matrix with each row of the first matrix (T_1) in the following way.

Consider multiplying the transformation matrix and the first column of the second matrix:

$$\begin{bmatrix} 1/2 & 0 \\ -3 & 1 \end{bmatrix} \begin{bmatrix} 2 \\ 6 \end{bmatrix} = \begin{bmatrix} 1/2 \times 2 & + & 0 \times 6 \\ -3 \times 2 & + & 1 \times 6 \end{bmatrix} = \begin{bmatrix} 1 \\ 0 \end{bmatrix}$$

For clarity the same computations are performed again, row by row of the transformation matrix:

$$[1/2 \quad 0] \begin{bmatrix} 2 \\ 6 \end{bmatrix} = [1/2 \times 2 + 0 \times 6] = 1$$

$$[-3 \quad 1] \begin{bmatrix} 2 \\ 6 \end{bmatrix} = [-3 \times 2 + 1 \times 6] = 0$$

Similar computations must be performed for each column of the original coefficient matrix. The result of premultiplying the entire matrix of coefficients by T_1 is given below:

$$
\begin{array}{cccc}
T_1 & x_1 & x_2 & \text{sol} \\
\begin{bmatrix} 1/2 & 0 \\ -3 & 1 \end{bmatrix} & \begin{bmatrix} \boxed{2} & 4 & 16 \\ 6 & 2 & 18 \end{bmatrix}
\end{array}
$$

$$
\begin{array}{ccc}
x_1 & x_2 & \text{sol} \\
\begin{bmatrix} 1 & 2 & 8 \\ 0 & -10 & -30 \end{bmatrix}
\end{array}
\quad \text{or} \quad
\begin{array}{rcrcr}
x_1 & + & 2x_2 & = & 8 \\
 & - & 10x_2 & = & -30
\end{array}
$$

The coefficients in the last matrix are identical to the coefficients of system (A*) in step 1 of the previous section. This is exactly the result we wanted to achieve. It is also important to realize that the coefficients of any tableau can be read as a system of equations by simply reattaching the corresponding variables.

The pivot method continues with a new transformation matrix T_2 corresponding to the procedures of step 2 in section 6.2. At each iteration the pivot must be chosen in rows and columns different from those previously selected. In this example, a choice no longer exists: the pivot must be (-10) in the second row and column of the last tableau.

$$
\begin{array}{cccc}
T_2 & x_1 & x_2 & \text{sol} \\
\begin{bmatrix} 1 & 2/10 \\ 0 & -1/10 \end{bmatrix} & \begin{bmatrix} 1 & 2 & 8 \\ 0 & \boxed{-10} & -30 \end{bmatrix}
\end{array}
$$

$$\begin{bmatrix} 1 & 0 & 2 \\ 0 & 1 & 3 \end{bmatrix}$$

Because the pivot (-10) was selected in row #2, the transformation vector in T_2 is placed in column #2. In the last tableau we have achieved the objective of transforming the original system of equations (A) into the form of system (B). This equivalence can be clearly seen by reattaching the variables to the corresponding coefficients of the final matrix as

$$
\begin{aligned}
x_1 &&= 2 &= \bar{x}_1 \\
x_2 &= 3 &= \bar{x}_2
\end{aligned}
$$

The solution of the pivot method is identical to that obtained with the Gaussian elimination procedure.

Verification of the solution. It is always very important to verify the solution obtained by checking that it satisfies the original equations as, indeed, it does in this case:

$$
\begin{aligned}
2(2) &+ 4(3) &= 16 \\
6(2) &+ 2(3) &= 18
\end{aligned}
$$

6.3.2 The Choice of a Pivot

In a system of linear equations the choice of a pivot is regulated by two simple rules:

| It must not be equal to zero. |

| At each iteration it must be chosen from different rows and columns. |

A pivot, therefore, can be chosen as an off-diagonal coefficient. To verify this freedom of selection and to understand its consequences, let us recompute the solution of system (A) starting with a different pivot, say coefficient 4 in equation #1.

$$
\begin{array}{c}
T_1 \\
\begin{bmatrix} 1/4 & 0 \\ -1/2 & 1 \end{bmatrix}
\end{array}
\begin{array}{ccc}
x_1 & x_2 & \text{sol} \\
\begin{bmatrix} 2 & \boxed{4} & 16 \\ 6 & 2 & 18 \end{bmatrix}
\end{array}
\qquad
\begin{array}{l}
\text{pivot} = 4 \text{ in row #1 col. #2} \\
\text{transf. vector in } T_1 \text{ is col. #1}
\end{array}
$$

$$
\begin{array}{c}
T_2 \\
\begin{bmatrix} 1 & -1/10 \\ 0 & 1/5 \end{bmatrix}
\end{array}
\begin{array}{ccc}
x_1 & x_2 & \text{sol} \\
\begin{bmatrix} 1/2 & 1 & 4 \\ \boxed{5} & 0 & 10 \end{bmatrix}
\end{array}
\qquad
\begin{array}{l}
\text{pivot} = 5 \text{ in row #2 col. #1} \\
\text{transf. vector in } T_2 \text{ is col. #2}
\end{array}
$$

$$
\begin{array}{ccc}
x_1 & x_2 & \text{sol} \\
\begin{bmatrix} 0 & 1 & 3 \\ 1 & 0 & 2 \end{bmatrix}
\end{array}
$$

By reattaching the variable names to the coefficients of the final tableau, we can easily read the solution as

$$x_2 = 3$$
$$x_1 = 2$$

which is the solution obtained before but with a different sequence of pivots. The only consequence of choosing pivots off the main diagonal of the original matrix is the permutation of rows in the final system.

6.3.3 Another Numerical Example

Let us gain confidence in the newly acquired knowledge by solving the following (3 × 3) system of equations using the pivot method:

$$
\begin{array}{rrrrrrrr}
4x_1 & + & 2x_2 & + & 3x_3 & = & 12 \\
-x_1 & - & 6x_2 & + & 2x_3 & = & 4 \\
2x_1 & + & 4x_2 & - & 2x_3 & = & 8
\end{array}
$$

Pivots will not be chosen on the main diagonal in order to reinforce the idea of freedom of pivot selection.

$$
\begin{array}{c}
T_1 \\
\begin{bmatrix} 1 & 0 & -2 \\ 0 & 1 & 1/2 \\ 0 & 0 & 1/2 \end{bmatrix}
\end{array}
\quad
\begin{array}{ccc}
x_1 & x_2 & x_3 & \text{sol} \\
\begin{bmatrix} 4 & 2 & 3 & 12 \\ -1 & -6 & 2 & 4 \\ \boxed{2} & 4 & -2 & 8 \end{bmatrix}
\end{array}
$$

pivot $= 2$ in row #3 col. #1 transf. vector in T_1 is col. #3

$$
\begin{array}{c}
T_2 \\
\begin{bmatrix} 1 & -7 & 0 \\ 0 & 1 & 0 \\ 0 & 1 & 1 \end{bmatrix}
\end{array}
\quad
\begin{array}{ccc}
x_1 & x_2 & x_3 & \text{sol} \\
\begin{bmatrix} 0 & -6 & 7 & -4 \\ 0 & -4 & \boxed{1} & 8 \\ 1 & 2 & -1 & 4 \end{bmatrix}
\end{array}
$$

pivot $= 1$ in row #2 col. #3 transf. vector in T_2 is col. #2

$$
\begin{array}{c}
T_3 \\
\begin{bmatrix} 1/22 & 0 & 0 \\ 4/22 & 1 & 0 \\ 2/22 & 0 & 1 \end{bmatrix}
\end{array}
\quad
\begin{array}{ccc}
x_1 & x_2 & x_3 & \text{sol} \\
\begin{bmatrix} 0 & \boxed{22} & 0 & -60 \\ 0 & -4 & 1 & 8 \\ 1 & -2 & 0 & 12 \end{bmatrix}
\end{array}
$$

pivot $= 22$ in row #1 col. #2 transf. vector in T_3 is col. #1

$$
\begin{array}{ccc}
x_1 & x_2 & x_3 & \text{sol} \\
\begin{bmatrix} 0 & 1 & 0 & -60/22 \\ 0 & 0 & 1 & -64/22 \\ 1 & 0 & 0 & 144/22 \end{bmatrix}
\begin{array}{l} = \bar{x}_2 \\ = \bar{x}_3 \\ = \bar{x}_1 \end{array}
\end{array}
$$

The solution of the given equations' system is $x_2 = -30/11$, $x_3 = -32/11$, and $x_1 = 72/11$. To solve a system of three equations, it was necessary to perform three transformations. In general, the number of iterations is equal to the number of equations.

The ability to solve systems of linear equations via the pivot method can be put to fruitful use in the solution of linear programming problems. This is so because an optimal solution always occurs at an extreme point, which, in turn, is associated with an appropriately specified basic system of equations. The difference between the pivot method for solving a system of equations and the simplex method for linear programming is that, in the latter, the choice of the pivot is not arbitrary but is guided by a profitability criterion.

6.4   The Inverse of a Basis

The solution of a system of linear equations requires the computation of the *inverse matrix* of coefficients of the original system. In solving the numerical examples presented above, the inverse matrix was computed but not displayed. The inverse matrix is of interest in its own way, but it also possesses an important economic interpretation, as will be indicated further on.

To gain familiarity with the notion of an inverse matrix, consider the following very simple equation in one variable:

$$3x = 6$$

In order to generalize the discussion, let $B = 3$ and $b = 6$. Then the above equation can be written as

(C) $$Bx = b$$

To solve for x in terms of the coefficients, we compute the reciprocal, or the inverse, of 3 and multiply both sides of the equation as follows:

$$\frac{1}{3}3x = \frac{1}{3}6$$
$$x = 3^{-1}6 = 2$$

In symbolic terms:

(D) $$\frac{1}{B}Bx = \frac{1}{B}b$$
$$x = B^{-1}b$$

Now reconsider the familiar system of equations (A):

(A)
$$\begin{array}{rcl} 2x_1 + 4x_2 & = & 16 \\ 6x_1 + 2x_2 & = & 18 \end{array}$$

which can be rewritten in matrix form as

(A)
$$\begin{bmatrix} 2 & 4 \\ 6 & 2 \end{bmatrix} \begin{bmatrix} x_1 \\ x_2 \end{bmatrix} = \begin{bmatrix} 16 \\ 18 \end{bmatrix}$$

By letting B stand for a basis,

$$B = \begin{bmatrix} 2 & 4 \\ 6 & 2 \end{bmatrix}, \ \mathbf{x} = \begin{bmatrix} x_1 \\ x_2 \end{bmatrix}, \ \text{and} \ \mathbf{b} = \begin{bmatrix} 16 \\ 18 \end{bmatrix}$$

it is possible to rewrite the (A) system in a very compact form equivalent to (C), $B\mathbf{x} = \mathbf{b}$.

The solution of this system can be obtained as in (D), that is,

(E)
$$\mathbf{x} = B^{-1}\mathbf{b}$$

The real difference between the expressions in (D) and in (E) is that in (D) all the components (B, x, and b) are scalars while in (E) they are vectors and matrices. The expression (E) applied to the numerical example is stated as

$$\begin{bmatrix} x_1 \\ x_2 \end{bmatrix} = \begin{bmatrix} 2 & 4 \\ 6 & 2 \end{bmatrix}^{-1} \begin{bmatrix} 16 \\ 18 \end{bmatrix}$$

and the matrix

$$B^{-1} = \begin{bmatrix} 2 & 4 \\ 6 & 2 \end{bmatrix}^{-1}$$

is the inverse of the basis

$$B = \begin{bmatrix} 2 & 4 \\ 6 & 2 \end{bmatrix}$$

From (D) it is clear that $B^{-1}B = 1$, while from (E) we have $B^{-1}B = I$, where I is the identity matrix. The matrix I is called the identity matrix because it is an extension of the number 1 on the real line. Its main role is to turn any number into itself, hence, the name of identity operator. In fact, given any real number a, $1 \times a = a$. In similar fashion, given any system of linear equations, the premultiplication (according to the rules explained in section 6.3.1) of the system by the identity matrix I leaves the system unchanged.

In order to find an explicit numerical representation of the inverse matrix B^{-1}, we need only to augment the tableau of the pivot method with an identity matrix as follows:

$$
\begin{array}{cccccc}
T_1 & & x_1 & x_2 & I & \text{sol} \\
\begin{bmatrix} 1/2 & 0 \\ -3 & 1 \end{bmatrix} &
\begin{bmatrix} \boxed{2} & 4 & \vdots & 1 & 0 & \vdots & 16 \\ 6 & 2 & \vdots & 0 & 1 & \vdots & 18 \end{bmatrix}
\end{array}
$$

$$
\begin{array}{cccccc}
T_2 & & x_1 & x_2 & T_1 I & \text{sol} \\
\begin{bmatrix} 1 & 2/10 \\ 0 & -1/10 \end{bmatrix} &
\begin{bmatrix} 1 & 2 & \vdots & 1/2 & 0 & \vdots & 8 \\ 0 & \boxed{-10} & \vdots & -3 & 1 & \vdots & -30 \end{bmatrix}
\end{array}
$$

$$
\begin{array}{cccc}
x_1 & x_2 & T_2 T_1 I & \text{sol} \\
\begin{bmatrix} 1 & 0 & \vdots & -1/10 & 2/10 & \vdots & 2 \\ 0 & 1 & \vdots & 3/10 & -1/10 & \vdots & 3 \end{bmatrix}
\end{array}
$$

The (2×2) matrix appearing in the final tableau between the dotted lines is the inverse of the basis

$$
B = \begin{bmatrix} 2 & 4 \\ 6 & 2 \end{bmatrix}
$$

Hence,

$$
B^{-1} = \begin{bmatrix} -1/10 & 2/10 \\ 3/10 & -1/10 \end{bmatrix}
$$

If the computations were correctly performed, $B^{-1}B = BB^{-1} = I$ and, indeed,

$$
\begin{bmatrix} -1/10 & 2/10 \\ 3/10 & -1/10 \end{bmatrix} \begin{bmatrix} 2 & 4 \\ 6 & 2 \end{bmatrix} = \begin{bmatrix} 2 & 4 \\ 6 & 2 \end{bmatrix} \begin{bmatrix} -1/10 & 2/10 \\ 3/10 & -1/10 \end{bmatrix} = \begin{bmatrix} 1 & 0 \\ 0 & 1 \end{bmatrix}
$$

Finally, notice another remarkable result: the premultiplication of all the transformation matrices (T_1 and T_2, in this example) is equal to the inverse matrix B^{-1}. In fact, $T_2 T_1 = B^{-1}$:

$$
\begin{bmatrix} 1 & 2/10 \\ 0 & -1/10 \end{bmatrix} \begin{bmatrix} 1/2 & 0 \\ -3 & 1 \end{bmatrix} = \begin{bmatrix} -1/10 & 2/10 \\ 3/10 & -1/10 \end{bmatrix}
$$

For this reason, this procedure for computing the inverse of a matrix is called the *product form of the inverse*.

6.5   A General View of the Pivot Method

During the computations of the pivot method the coefficients of the equation system assume an interesting structure that is revealed only by dealing with a symbolic example such as

$$a_{11}x_1 + a_{12}x_2 = b_1$$
$$a_{21}x_1 + a_{22}x_2 = b_2$$

Definition of the initial tableau and of the transformation matrix T_1. In the following example, the pivot is chosen as the coefficient a_{12} in row #1 and the transformation vector replaces column 1 in T_1.

$$
\begin{array}{cc}
T_1 & \\
\begin{bmatrix} \dfrac{1}{a_{12}} & 0 \\[2ex] -\dfrac{a_{22}}{a_{12}} & 1 \end{bmatrix}
&
\begin{array}{ccc}
x_1 & x_2 & \text{sol} \\
\end{array} \\
& \begin{bmatrix} a_{11} & \boxed{a_{12}} & b_1 \\[1ex] a_{21} & a_{22} & b_2 \end{bmatrix}
\end{array}
$$

$$
\begin{array}{ccc}
x_1 & x_2 & \text{sol} \\
\end{array}
$$
$$
\begin{bmatrix}
\dfrac{a_{11}}{a_{12}} & \dfrac{a_{12}}{a_{12}} & \dfrac{b_1}{a_{12}} \\[2ex]
\left\{ a_{21} - \dfrac{a_{22}}{a_{12}}a_{11} \right\} & \left\{ a_{22} - \dfrac{a_{22}}{a_{12}}a_{12} \right\} & \left\{ b_2 - \dfrac{a_{22}}{a_{12}}b_1 \right\}
\end{bmatrix}
$$

Definition of an intermediate tableau and the transformation matrix T_2. In order to proceed to a second iteration, we must rewrite the above tableau in a more manageable form. The second column of coefficients is equal to a unit vector. Therefore, let

$$a_{11}^* \equiv \frac{a_{11}}{a_{12}}, \quad a_{21}^* \equiv a_{21} - \frac{a_{22}}{a_{12}}a_{11}$$

$$b_1^* \equiv \frac{b_1}{a_{12}}, \quad b_2^* \equiv b_2 - \frac{a_{22}}{a_{12}}b_1$$

The second iteration begins from the final tableau of the previous step and with the definition of the transformation matrix T_2, selecting a pivot in different rows and columns from those of previous pivots.

$$T_2 \qquad\qquad x_1 \quad\; x_2 \quad\;\; \text{sol}$$

$$\begin{bmatrix} 1 & -\dfrac{a_{11}^*}{a_{21}^*} \\[2ex] 0 & \dfrac{1}{a_{21}^*} \end{bmatrix} \begin{bmatrix} a_{11}^* & 1 & b_1^* \\[1ex] \boxed{a_{21}^*} & 0 & b_2^* \end{bmatrix}$$

$$\qquad\qquad x_1 \quad\; x_2 \qquad\quad \text{sol}$$

$$\begin{bmatrix} 0 & 1 & \left\{b_1^* - \dfrac{a_{11}^*}{a_{21}^*} b_2^*\right\} \\[3ex] 1 & 0 & \dfrac{b_2^*}{a_{21}^*} \end{bmatrix}$$

The starred coefficients in this second iteration exhibit the apparent structure of the original coefficients in the first iteration. Their inner structure, however, is more elaborate, exhibiting the symmetric recursion framework that is at the foundation of the pivot algorithm:

$$\frac{a_{11}^*}{a_{21}^*} = \frac{\dfrac{a_{11}}{a_{12}}}{a_{21} - \dfrac{a_{22}}{a_{12}} a_{11}}$$

$$\frac{b_2^*}{a_{21}^*} = \frac{b_2 - \dfrac{a_{22}}{a_{12}} b_1}{a_{21} - \dfrac{a_{22}}{a_{12}} a_{11}}$$

$$b_1^* - \frac{a_{11}^*}{a_{21}^*} b_2^* = \frac{b_1}{a_{12}} - \frac{\dfrac{a_{11}}{a_{12}}}{a_{21} - \dfrac{a_{22}}{a_{12}} a_{11}} \left(b_2 - \frac{a_{22}}{a_{12}} b_1 \right)$$

As the number of iterations increases, the apparent complexity of the computations increases but the shape of the coefficients remains the same, revealing the fractal nature of the pivot algorithm.

Chapter 7

The Primal Simplex Algorithm

7.1 Introduction to LP Algorithms

Since the discovery of the simplex method in 1947, many different approaches have been proposed for solving linear programming problems. Surprisingly, the first method still seems to be the best. The simplex method is "best" in the sense that it is a very efficient procedure for solving a large class of problems and, furthermore, every single step can be given an economic interpretation.

We are ready to study procedures, or algorithms, for solving linear programming problems in a very efficient way. The need for more than one algorithm is based upon the dual nature of linear programming. Stunningly, duality and symmetry also extend their effect to algorithms!

As the title of this chapter suggests, if there is an algorithm called the *primal simplex method,* another one called the *dual simplex method* must also exist. Indeed, such an algorithm does exist: symmetry applies not only to problems but also to algorithms.

Let us reflect for a moment. Problems can be likened to cars (and cars to problems). A dual pair of linear programming problems is similar to two beautiful cars: a Ferrari and a Lamborghini. Mr. Ferrari and Mr. Lamborghini lived in the same town of Modena in northern Italy. The story goes that Mr. Lamborghini, a manufacturer of tractors, used to drive a Ferrari car but kept complaining to Mr. Ferrari that the car was always breaking down. One frustrating day, Mr. Lamborghini told Mr. Ferrari in no uncertain terms that he was not able to build a decent car. "You should not be allowed to speak about the subject," Mr. Ferrari replied, "since you are able to build only tractors. Maybe you drive my Ferrari as you drive your tractors!" The legend concludes with the following epilogue: Mr. Lamborghini could not take the insult lightly and no longer drove a Ferrari. Instead, he went on to build cars to show his friend Enzo how wrong he was. And the Lamborghini car was born.

Let us agree, therefore, to consider the Ferrari as the primal problem while the Lamborghini is the dual problem. These two cars (problems) come with their own sets of tools. We can liken the two toolboxes to the algorithms for solving the car problems. Similarly, the primal simplex and the dual simplex methods are the algorithms (toolboxes) for solving the primal and the dual problems.

It is not inconceivable that the Ferrari toolbox could be used to fix (solve) the Lamborghini car (problem) and vice versa. Analogously, as we will soon find out, it is possible to use the primal simplex algorithm for solving both primal and dual problems. Conversely, it is possible to use the dual simplex algorithm to solve dual and primal problems.

The story of the two simplex algorithms is very interesting. The name of G. B. Dantzig is usually associated with the discovery of the primal simplex method in 1947, although he amply acknowledged the contribution of several other scientists. The intellectual process of trial and error Dantzig and associates went through before recognizing the potential of the simplex method is a story in itself. Seven years after the appearance of the primal simplex algorithm, C. E. Lemke discovered the dual simplex method in 1954, as an application of the primal simplex method to the dual problem. The history of numerical problem solving has never been the same since.

The overall picture of the linear programming algorithms is not complete without introducing another dual pair of auxiliary procedures. The reason is as follows: because both primal and dual simplex algorithms require beginning computations with an explicit basic feasible solution, the question of what to do when feasibility is not easily discernible has induced the invention of the artificial variable and the artificial constraint algorithms. As discussed in chapter 2, the dual of a variable is a constraint and the dual of a constraint is a variable. Hence, the new set of procedures really constitutes a dual pair of algorithms.

The objective of the four algorithms for solving linear programming problems can be summarized in table 7.1.

The algorithms of table 7.1 will be studied in the following order:

1. Primal simplex algorithm.
2. Dual simplex algorithm.
3. Artificial variable algorithm.
4. Artificial constraint algorithm.

Table 7.1. Objectives of LP Simplex Algorithms

	Primal Method		Dual Method	
	Objective	Algorithm	Objective	Algorithm
Phase I	obtain primal feasibility	←⊢ artificial variable	obtain dual feasibility	←⊢ artificial constraint
Phase II	maintain primal feasibility; achieve primal optimality	←⊢ primal simplex	maintain dual feasibility; achieve dual optimality	←⊢ dual simplex

This order is dictated by the desire to solve a linear programming problem as soon as possible. The first three algorithms are essential: armed with this triad, we will be able to solve any linear programming problem. The fourth algorithm is presented for completing the symmetry and for deeper understanding of the working of the other algorithms.

7.2   The Primal Simplex Algorithm

Mathematically, a simplex is a triangle in a 2-dimensional space, a tetrahedron (a four-edged object) in a 3-dimensional space, a pentahedron (a five-edged object) in a 4-dimensional space, and so forth. At each iteration, the primal simplex algorithm constructs a simplex in the appropriate space; hence, the name of the algorithm.

Before starting the presentation of the algorithm, we review the knowledge about LP problems acquired thus far. A primal problem such as

Primal maximize Z $= c_1 x_1 + c_2 x_2$

subject to $\quad a_{11} x_1 + a_{12} x_2 \leq b_1$
$\qquad\qquad\quad a_{21} x_1 + a_{22} x_2 \leq b_2$

$$x_1 \geq 0, \ x_2 \geq 0$$

can be dissected by means of the diagrammatic scheme in figure 7.1.

The scheme in figure 7.1 illustrates the chain of reasoning that lies at the foundation of the simplex algorithm. When the linear programming problem has the dimensions of the above example, a primal system of constraints can be represented in both the output space and the input requirement space. From the output space, it is easy to identify the extreme points of the feasible region (in qualitative terms). In turns, from

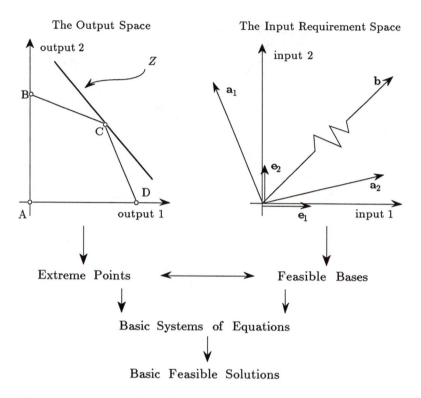

Figure 7.1. The flow chart for the primal simplex method.

the input requirement space, it is possible to list all the feasible bases of the primal problem, including those that contain slack activities, even though such activities were not explicitly stated in the above primal problem. A feasible basis is a basis that forms a cone of less than 180°, which includes the right-hand-side vector of constraint coefficients **b**. Feasible bases and extreme points are related by a one-to-one correspondence (in nondegenerate problems). From the qualitative knowledge of an extreme point, it is possible to reduce a primal system of inequality constraints to a basic system of equations whose solution will provide an extreme point in quantitative terms. Such a representation of the extreme point is called a basic feasible solution. Equivalently, knowledge of the vectors forming a feasible basis derived from the input requirement space is sufficient for reducing the system of inequality constraints to a basic system of equations and, therefore, for obtaining a basic feasible solution. The importance of the above scheme is that it allows a thorough understanding of the linear programming structure using two-dimensional diagrams. After achieving this objective, it is a simple matter to generalize the notions of feasible basis and of basic feasible solution to higher dimensional problems without the helping graphs.

7.2.1 Ideas Underlying the Primal Simplex Method

Step 1. Since the optimal value of the objective function occurs at an extreme point (see the output space of figure 7.1), the initial idea is to explore only extreme points. We know from a previous discussion that extreme points are associated with feasible bases, which in turn produce basic feasible solutions. Therefore, the rule for this step is:

> Choose a feasible basis and compute the corresponding basic feasible solution.

Step 2. The rule for the second step is:

> Verify whether the basic feasible solution obtained in step 1 is also an optimal solution.

If it is, stop. If not, continue.

Step 3. In the third step, the rule is:

> Select an adjacent extreme point (a new basic feasible solution) by changing the old feasible basis by only one column vector.

Go to step 2.

The ideas presented in steps 1, 2, and 3 are relatively simple and follow directly from the analysis developed in previous chapters. Efficiency and simplicity explain the extraordinary success of the simplex method. The attribute "simplex" associated with the algorithm, however, is not related in any way to the idea of simplicity but to the geometric objects introduced at the beginning of section 7.2.

From step 1, the primal simplex method requires the availability of an explicit primal basic feasible solution. A cycle composed of steps 2 and 3 is called an iteration. The implementation of steps 1, 2, and 3 will be illustrated with reference to the following example:

$$(P) \quad \text{maximize } Z = \quad 3x_1 + \quad 5x_2$$

$$\text{subject to} \quad 2x_1 + \quad 4x_2 \leq 16 \leftarrow \text{land}$$
$$6x_1 + \quad 3x_2 \leq 18 \leftarrow \text{labor}$$

$$\underset{\text{wheat}}{\uparrow} \qquad \underset{\text{corn}}{\uparrow}$$

and x_1, $x_2 \geq 0$. All constraints that represent inequalities admit slack variables. As in previous examples, therefore, it is convenient to reformulate the above LP problem (P) into an equivalent one (P') where all inequality constraints have become equations with the introduction of nonnegative slack variables and all constant terms in the problem appear in an orderly fashion on one side of the equality sign:

$$(P') \quad \text{maximize } Z - 3x_1 - 5x_2 - 0x_{s1} - 0x_{s2} = 0$$

$$\text{subject to} \quad \begin{aligned} 2x_1 + 4x_2 + x_{s1} \quad\quad &= 16 \\ 6x_1 + 3x_2 \quad\quad + x_{s2} &= 18 \end{aligned}$$

At this stage, the diagrammatic scheme of figure 7.1 suggests the graphing of the output and input requirement spaces as in figures 7.2a and 7.2b. This task can be better accomplished by reformulating the system of primal constraints in terms of activity vectors:

$$\underset{\mathbf{a_1}}{\begin{bmatrix} 2 \\ 6 \end{bmatrix}} x_1 + \underset{\mathbf{a_2}}{\begin{bmatrix} 4 \\ 3 \end{bmatrix}} x_2 + \underset{\mathbf{e_1}}{\begin{bmatrix} 1 \\ 0 \end{bmatrix}} x_{s1} + \underset{\mathbf{e_2}}{\begin{bmatrix} 0 \\ 1 \end{bmatrix}} x_{s2} = \underset{\mathbf{b}}{\begin{bmatrix} 16 \\ 18 \end{bmatrix}}$$

Iteration #1:

Step 1. The primal simplex algorithm requires the availability of an explicit basic feasible solution. This initial condition implies that a feasible basis should be at hand. In the above numerical example, feasible bases can be determined by a simple inspection of figure 7.2b. The idea is, therefore, to start the primal simplex algorithm by selecting the easiest feasible basis and computing the associated basic feasible solution. The easiest feasible basis is, of course, the identity basis I, which is entirely composed of slack vectors, that is,

$$\text{easiest feasible basis } = I = \begin{bmatrix} 1 & 0 \\ 0 & 1 \end{bmatrix} \quad \text{(from figure 7.2b)}$$

to which there corresponds the associated

$$\text{extreme point } A = \{x_1 = 0, x_2 = 0, x_{s1} > 0, x_{s2} > 0\}$$

as determined from figure 7.2a. From the graphical identification of either the feasible basis I or the extreme point A, it is a simple matter to set up (from problem P') the associated basic system of equations whose solution will provide the initial basic feasible solution:

$$\text{basic system of equations :} \quad \begin{bmatrix} 1 & 0 \\ 0 & 1 \end{bmatrix} \begin{bmatrix} x_{s1} \\ x_{s2} \end{bmatrix} = \begin{bmatrix} 16 \\ 18 \end{bmatrix}$$
$$\mathbf{e_1} x_{s1} + \mathbf{e_2} x_{s2} = \mathbf{b}$$

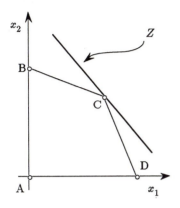 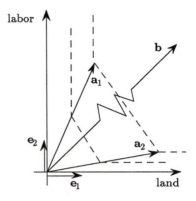

Figure 7.2. (a) Output space.   (b) Input requirement space.

The resulting basic feasible solution is found by inspection of the above system of equations with $x_{s1} = 16$ and $x_{s2} = 18$, plus the other components of extreme point A, x_1, and x_2, which we already know are equal to zero. The graphical representation of the initial feasible basis and basic feasible solution is given in figure 7.3.

Viewed as in figure 7.3, a basic feasible solution $(x_1 = 0, x_2 = 0, x_{s1} = 16, x_{s2} = 18)$ is a set of scalars that stretch (or shrink) the associated basis vectors in such a way as to form a parallelogram whose main diagonal is the vector of constraint coefficients **b.**

Step 2. After finding the current basic feasible solution, it is necessary to decide whether it is an optimal solution. This is done by evaluating the current basic feasible solution in terms of the coefficients of the objective function, using the specification of problem (P′):

$$
\begin{aligned}
Z_A &= 3x_1 + 5x_2 + 0x_{s1} + 0x_{s2} \\
Z_A &= 0
\end{aligned}
$$

Is this initial basic feasible solution optimal? The answer is no because, according to problem (P′), if either $x_1 > 0$ or $x_2 > 0$, then $Z > Z_A = 0$. The recommended action at this stage is, therefore, to change the current feasible basis by replacing a slack activity vector with either the wheat or the corn activity vector. More precisely, the entry criterion is formulated in the following general form, keeping in mind the specification of the linear

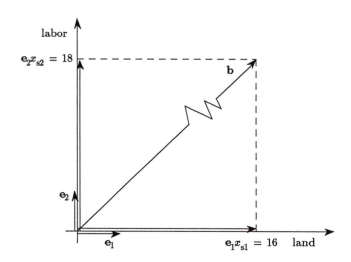

Figure 7.3. The initial feasible basis and basic feasible solution.

programming problem as in (P'):

> **Entry criterion:** To improve the current feasible solution, select the activity associated with the most negative coefficient in the objective function.

In the example under discussion, the corn activity, a_2, satisfies the entry criterion with $-c_2 = -5 < -c_1 = -3$. Hence, we expect $x_2 > 0$ in the new basic feasible solution.

Step 3. A new basis is constructed from the old (current) basis by changing only one vector. This operation shifts the basic feasible solution from one extreme point to an adjacent one. Activity a_2 was selected in the previous step for entering the basis. In order to complete the change of basis, the following question must be answered: Which activity (column) should be eliminated from the current feasible basis? The question possesses economic meaning. After all, the land and labor input availabilities are presently exhausted among slack activities e_1 and e_2. The answer, therefore, must take into account the fact that the new basis must also be feasible; that is, it must return a new basic feasible solution corresponding to a producible plan. In other words, all the components of the solution vector x, including the slack variables, must be nonnegative and satisfy the input constraints.

To summarize our knowledge of the problem accumulated up to this point, we list the current basic feasible solution and the desired change $x_2 > 0$, together with its consequences:

<table>
<tr><td colspan="2" align="center">current basic
feasible solution
(extreme point A)</td><td colspan="2" align="center">new basic
feasible solution
prior to discussion</td></tr>
<tr><td>x_1</td><td>$=\quad 0$</td><td>x_1</td><td>$=\quad 0$</td></tr>
<tr><td>x_2</td><td>$=\quad 0$</td><td>x_2</td><td>$>\quad 0$</td></tr>
<tr><td>x_{s1}</td><td>$=\quad 16$</td><td>x_{s1}</td><td>$=\quad 16 - 4x_2 \geq 0$</td></tr>
<tr><td>x_{s2}</td><td>$=\quad 18$</td><td>x_{s2}</td><td>$=\quad 18 - 3x_2 \geq 0$</td></tr>
</table>

Discussion. The desired increase in the level of x_2, as determined from step 2, requires a decrease in x_{s1} and x_{s2} by an amount dictated by the coefficients of the corn activity. In other words, each unit of corn requires 4 units of land and 3 units of labor. Currently, all the available land and labor are employed by the slack activities, which, therefore, must release some resources to allow the cultivation of corn. To preserve the feasibility of the new basic feasible solution, x_2 must be chosen in such a way as to maintain x_{s1} and x_{s2} nonnegative. This objective is achieved only by choosing the level of x_2 as follows:

$$16 \geq 4x_2 \qquad \rightarrow x_2 = \frac{16}{4} = 4 \qquad \text{from } x_{s1}$$

$$18 \geq 3x_2 \qquad \rightarrow x_2 = \frac{18}{3} = 6 \qquad \text{from } x_{s2}$$

The conclusion is that the increase in the level of activity #2 (corn) should be in the amount of

$$x_2 = \min \left\{ \frac{16}{4}, \frac{18}{3} \right\} = \min\{4, 6\} = 4$$

where min stands for minimum. The minimum ratio criterion developed here is a necessary condition for maintaining the feasibility of the new basic solution, which, after the discussion, can be stated as

new basic feasible solution after discussion
(extreme point B, figure 7.2a)

$$
\begin{aligned}
x_1 &= 0 \\
x_2 &= 4 \\
x_{s1} &= 16 - 4(4) = 0 \\
x_{s2} &= 18 - 3(4) = 6
\end{aligned}
$$

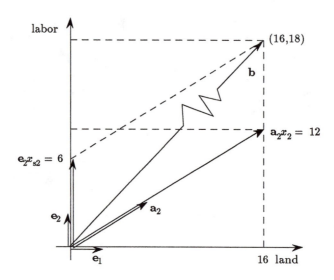

Figure 7.4. Representation of the second feasible basis and basic feasible solution.

Notice that by choosing the level of x_2 without following the minimum ratio criterion, the value of x_{s1} would be negative, $x_{s1} = 16 - 4(6) = -8$, violating the feasibility of the basic solution. Furthermore, feasibility of the new solution is maintained only by using positive coefficients in the denominators of the ratios. The criterion for exiting the old feasible basis can thus be stated in general terms as

> **Exit criterion:** To change the current feasible basis, eliminate the column corresponding to the index in the numerator of the minimum ratio.

The new activity, chosen in step 2, will increase by the amount corresponding to the minimum ratio, that is,

$$x_{\text{new}} = \min \left\{ \frac{\text{values of the current feasible solution}}{\text{positive coefficients of the new activity}} \right\}$$

In the above numerical example, the minimum ratio occurred in correspondence with the slack variable x_{s1}, which was reduced to zero in the new basic feasible solution. The column (activity) associated with x_{s1} is e_1, which will be eliminated from the current feasible basis. Its place will be taken up by the new activity vector a_2 associated with the production of corn. The new feasible basis is, therefore,

$$\text{new feasible basis} = \begin{matrix} x_2 & x_{s2} \\ \begin{bmatrix} 4 & 0 \\ 3 & 1 \end{bmatrix} \end{matrix}$$

Notice that the new basic feasible solution computed above is consistent with this new feasible basis, since it verifies the system of basic equations:

$$\begin{bmatrix} 4 & 0 \\ 3 & 1 \end{bmatrix} \begin{bmatrix} x_2 & = & 4 \\ x_{s2} & = & 6 \end{bmatrix} = \begin{bmatrix} 16 \\ 18 \end{bmatrix}$$

The second basic feasible solution, just computed and corresponding to extreme point B in figure 7.2a, can be graphed as in figure 7.4. The value of $x_2 = 4$ is the scalar that stretches activity vector

$$\mathbf{a}_2 = \begin{bmatrix} 4 \\ 3 \end{bmatrix}$$

to the point where the perpendicular line from the tip of the **b** vector intersects the extension of $\mathbf{a}_2$. Similarly, $x_{s2} = 6$ is the scalar that stretches activity vector

$$\mathbf{e}_2 = \begin{bmatrix} 0 \\ 1 \end{bmatrix}$$

to the point where it intersects the parallel to $\mathbf{a}_2$. The old basis is represented in figure 7.4 for the purpose of assessing the geometric modifications intervening in a change of basis procedure.

It remains to check whether the current feasible basis is optimal, a process that requires returning to step 2.

Iteration #2:

Step 2. The task is to evaluate the current basic feasible solution in terms of the objective function coefficients. Let us recall that the current basic feasible solution $\{x_1 = 0,\ x_2 = 4,\ x_{s1} = 0, x_{s2} = 6\}$ corresponds to extreme point B in figure 7.2a, which is associated with an objective function value equal to

$$\begin{aligned} Z_B &= 3x_1 + 5x_2 + 0x_{s1} + 0x_{s2} \\ Z_B &= 3(0) + 5(4) + 0(0) + 0(6) = 20 \end{aligned}$$

In going from extreme point A to B, the objective function value has increased from 0 to 20. The question is whether it is possible to improve it further. In other words, now that all the available land is allocated for the cultivation of corn, with six units of labor unemployed, can we find a new combination of productive activities such that Z is greater than 20?

In our simplified example, only wheat (x_1) is a candidate for a possible improvement of the production plan. In order to produce wheat, however, we must reduce the acreage of corn to which all the available land was allocated during the first iteration. The reduction of corn by one unit of output releases four units of land, which can then be used by wheat in a relatively more efficient way. Relative efficiency is measured here only in physical (technical) terms. According to this criterion, one unit of wheat employs two units of land, while one unit of corn employs four units of land. We thus say that judging by the input land alone, wheat is relatively more efficient than corn by the ratio of 2/4 or 1/2. Technical efficiency is a necessary condition, but it is not the ultimate criterion for judging efficiency. Economic efficiency (profitability) is what counts in the last analysis.

Therefore, to determine whether it is convenient to redirect resources from corn to wheat, we must compute the *opportunity cost* of wheat in terms of corn, a notion that requires the concept of *marginal rate of technical transformation.*

7.2.2   The Marginal Rate of Technical Transformation

Consider the following *transformation function* between two commodities x_1 and x_2:

$$f(x_1, x_2) = b$$

where b is the fixed quantity of a third commodity. In economics, the notion of transformation function is introduced in the discussion of a tradeoff between two commodities: they are goods in consumer theory and outputs and inputs in the theory of the firm. These various cases can be graphically represented by maps of level curves as in figure 7.5.

In all the diagrams of figure 7.5, the *marginal rate of technical transformation* $(MRTT)$ is defined as the ratio that measures how much we must decrease (or increase) x_1 to obtain a unit increase of x_2. In the limit, this ratio is given by the slope of the tangent line (or plane, in higher dimensions) to the function $f(x_1, x_2) = b$ or, equivalently, by the derivative $\partial x_2 / \partial x_1$. To compute such a derivative, we first take the total differential of $f(x_1, x_2) = b$ as

$$\frac{\partial f}{\partial x_1} dx_1 + \frac{\partial f}{\partial x_2} dx_2 = 0$$

and rearrange it to obtain the marginal rate of technical transformation:

$$MRTT_{x_2, x_1} \equiv -\frac{\partial x_2}{\partial x_1} = \frac{\partial f / \partial x_1}{\partial f / \partial x_2}$$

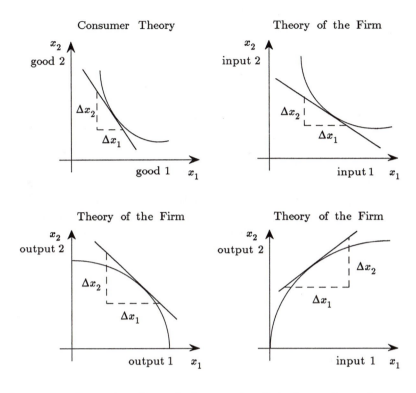

Figure 7.5. Examples of marginal rates of technical transformation.

In economics textbooks the above relation has different names, depending on the types of commodities involved. If the two commodities are inputs, the above ratio of derivatives is referred to as the marginal rate of technical substitution. If the commodities are an output and an input, we deal with marginal products. Finally, if the commodities are both outputs, the terminology is that of a marginal rate of technical transformation. In this book we will use this last terminology in any of the three situations, since in a linear programming problem the same commodity can be regarded as both an input and an output. As a consequence, the *MRTT* can be either positive, or zero, or negative.

Obviously, the notion of *MRTT* applies to linear functions such as the constraints in the linear programming problem (P′). Thus, consider the first constraint of that problem, which is rewritten here for convenience:

$$2x_1 + 4x_2 + x_{s1} = 16$$

The total differential of this linear function is

$$2dx_1 + 4dx_2 + dx_{s1} = 0$$

and the *MRTT* between x_2 and x_1 (with x_{s1} considered as a constant) is

$$MRTT_{x_2,x_1} \equiv -\frac{\partial x_2}{\partial x_1} = 4^{-1}2 = \frac{1}{2}$$

As mentioned before, linear programming is of interest (among other things) because it can handle multiple outputs and inputs simultaneously. It is appropriate, therefore, to extend the notion of marginal rate of technical transformation to several outputs and inputs. For illustration purposes, the numerical linear programming example formulated in problem (P′) will suffice. The constraints of that problem are rewritten here:

$$
\begin{array}{ccccccccll}
2x_1 & + & 4x_2 & + & x_{s1} & & & = & 16 & \leftarrow \text{ land} \\
6x_1 & + & 3x_2 & & & + & x_{s2} & = & 18 & \leftarrow \text{ labor} \\
\uparrow & & \uparrow & & \uparrow & & \uparrow & & & \\
\text{output} & & \text{output} & & \text{input} & & \text{input} & & &
\end{array}
$$

and, grouping the commodities in an arbitrary way, they can be restated in matrix form as

$$
\underbrace{\begin{bmatrix} 2 & 1 \\ 6 & 0 \end{bmatrix} \begin{bmatrix} x_1 \\ x_{s1} \end{bmatrix}}_{\substack{\uparrow \\ A\mathbf{x}_1}} + \underbrace{\begin{bmatrix} 4 & 0 \\ 3 & 1 \end{bmatrix} \begin{bmatrix} x_2 \\ x_{s2} \end{bmatrix}}_{\substack{\uparrow \\ B\mathbf{x}_2}} = \underset{\substack{\uparrow \\ \mathbf{b}}}{\begin{bmatrix} 16 \\ 18 \end{bmatrix}}
$$

where

$$
A = \begin{bmatrix} 2 & 1 \\ 6 & 0 \end{bmatrix}, \ B = \begin{bmatrix} 4 & 0 \\ 3 & 1 \end{bmatrix}, \ \mathbf{x}_1 = \begin{bmatrix} x_1 \\ x_{s1} \end{bmatrix}, \ \mathbf{x}_2 = \begin{bmatrix} x_2 \\ x_{s2} \end{bmatrix}, \ \mathbf{b} = \begin{bmatrix} 16 \\ 18 \end{bmatrix}
$$

The formulation of the constraints in matrix form allows for the conceptual treatment of the entire system as if we were dealing with only two commodities and one equation. The marginal rate of technical transformation between vectors can then be indicated as above by first taking the total differential of the matrix system:

$$A d\mathbf{x}_1 + B d\mathbf{x}_2 = 0$$

and then defining the *MRTT* in the usual way as

$$MRTT_{\mathbf{x}_2,\mathbf{x}_1} \equiv -\frac{\partial \mathbf{x}_2}{\partial \mathbf{x}_1} = B^{-1}A$$

This *MRTT*, which relates to vectors of commodities, has the same structural form as the *MRTT* involving only two commodities and, therefore, retains the economic meaning elaborated for the simple case. One word of caution is in order: since B^{-1} is the inverse matrix of B, it is improper and meaningless to write $1/B$. The marginal rate of technical transformation between vectors of commodities is a powerful way to compute and interpret the corresponding relations among many outputs and inputs simultaneously.

7.2.3 The Opportunity Cost of a Commodity

The *opportunity cost* of a commodity is the sacrifice of producing one additional unit of that commodity, measured in terms of alternative production opportunities that must be foregone. The sacrifice can be either positive or negative. A positive sacrifice is to be avoided, while a negative sacrifice is welcome. With this general definition, the opportunity cost of wheat can be stated as follows:

$$
\begin{matrix} \text{opportunity cost} \\ \text{of wheat } (x_1) \end{matrix} = \begin{pmatrix} \text{sacrifice} \\ \text{in terms} \\ \text{of corn} \end{pmatrix} - \begin{pmatrix} \text{unit} \\ \text{revenue} \\ \text{of wheat} \end{pmatrix}
$$

$$
= \frac{\text{wheat land}}{\text{corn land}} \begin{pmatrix} \text{unit revenue} \\ \text{of corn} \end{pmatrix} - \begin{pmatrix} \text{unit revenue} \\ \text{of wheat} \end{pmatrix}
$$

$$
= \frac{2}{4}5 \qquad - \quad 3 \quad = \quad -\frac{1}{2}
$$

Wheat land and corn land stand for the land coefficients of wheat and corn, respectively. The ratio $(2/4)$ is the marginal rate of technical transformation between wheat and corn measured in land units, as computed in the previous section. The unit revenue of corn is \$5, while that of wheat is \$3. At this stage, only the land constraint is binding, or in other words, none of the available land is allocated to the slack activity. For this reason the *MRTT* is measured solely in terms of land. The land released from one unit of corn (foregone to allow operating wheat at positive levels) can be better used by wheat with a technical efficiency ratio of 2 to 4. As a consequence, the loss of \$5 from one unit of corn foregone is mitigated by the technically more efficient wheat and, even though this activity brings in only \$3 per unit of output, the opportunity cost of wheat is negative. It is convenient, therefore, to redistribute the available resources from corn to wheat because this sacrifice is negative.

The notion of opportunity cost provides the general way for implementing the entry criterion. From now on, the definition of entry criterion given above must be interpreted as referring to coefficients of the objective function, which are opportunity costs of the corresponding activities.

We are ready to complete the execution of step 2 in iteration #2. By following the entry criterion redefined in terms of opportunity costs, it is convenient to operate the wheat activity at positive levels, $x_1 > 0$, and activity #1, a_1, should be brought into the new feasible basis.

Step 3. To determine the activity (column) that must leave the current feasible basis, the following observation is useful. In order to produce wheat, resources from corn must be released according to the *MRTT*. At this stage, land is the only limiting resource. Hence, the original input requirements of corn must be adjusted to reflect the relative technical advantage (in terms of land) that wheat exhibits over corn. This adjustment is analogous to the opportunity cost for the coefficients of the objective function. In other words, now that at least one resource (land) is exhausted, input requirements for an activity (wheat) that is a candidate for entering the new feasible basis must be measured in relation to that limiting resource. To achieve this objective, it is convenient to introduce the notion of *opportunity input requirement,* analogous to the concept of opportunity cost. The opportunity input requirement of a given commodity (wheat) is the savings (as opposed to sacrifice) of inputs attributable to one unit of a foregone activity (corn) adjusted by the marginal rate of technical transformation. Using the above numerical example,

$$\begin{pmatrix} \text{opportunity labor} \\ \text{requirement} \\ \text{for wheat} \end{pmatrix} = \begin{pmatrix} \text{wheat} \\ \text{labor} \\ \text{requirement} \end{pmatrix} - \begin{pmatrix} \text{savings} \\ \text{in terms} \\ \text{of corn} \end{pmatrix}$$

$$= \quad 6 \quad - \quad \frac{1}{2} \times 3 = \frac{9}{2}$$

For the input land, the adjustment is given by the marginal rate of technical transformation which, in this case, is equal to 1/2. These figures are the current labor and land requirements for wheat. They have been computed keeping in mind the relative efficiency (*MRTT*) with which wheat uses the limiting input land in comparison to corn. Analogous to the interpretation of opportunity cost, the opportunity input requirement measures, in physical terms, the advantage or disadvantage of shifting limiting resources from one activity to another.

The immediate task now is to decide which of the activities in the current feasible basis (corn and unused labor) should be replaced by wheat.

The second task is to determine at what level. We must repeat here the same type of discussion about current and new basic feasible solutions as in iteration #1. Starting with the current basic feasible solution (BFS), wheat was found to be profitable and, therefore, ought to be cultivated ($x_1 > 0$); the discussion will establish at what level.

current BFS	new BFS before discussion	new BFS after discussion
$x_1 = 0$	$x_1 > 0$	$x_1 = \dfrac{4}{3}$
$x_2 = 4$	$x_2 = 4 - \dfrac{1}{2}x_1 \geq 0$	$x_2 = \dfrac{10}{3}$
$x_{s1} = 0$	$x_{s1} = 0$	$x_{s1} = 0$
$x_{s2} = 6$	$x_{s2} = 6 - \dfrac{9}{2}x_1 \geq 0$	$x_{s2} = 0$

The new basic feasible solution corresponds to extreme point C in figure 7.2a. The transition from the new basic feasible solution before discussion to the new basic feasible solution after discussion involves the computation of the minimum ratio, which requires positive coefficients as denominators:

$$\text{from eq. of } x_2 \ : \ 4 \geq \frac{1}{2}x_1 \ \rightarrow \ x_1 = \frac{4}{1/2} = \frac{x_2}{1/2} = 8$$

$$\text{from eq. of } x_{s2} \ : \ 6 \geq \frac{9}{2}x_1 \ \rightarrow \ x_1 = \frac{6}{9/2} = \frac{x_{s2}}{9/2} = \frac{4}{3}$$

Therefore, the exit criterion for determining the activity that must leave the current feasible basis is

$$x_1 = \min\left\{ \frac{4}{1/2}, \frac{6}{9/2} \right\} = \min\left\{ 8, \frac{4}{3} \right\} = \frac{4}{3} = \frac{x_{s2}}{9/2}$$

The activity surplus labor, whose level appears in the numerator of the minimum ratio, leaves the current feasible basis. The new feasible basis (FB) is

$$\text{new } FB = \begin{array}{cc} x_2 & x_1 \\ \left[\begin{array}{cc} 4 & 2 \\ 3 & 6 \end{array} \right] \end{array} \quad \text{(keep the replacement order of the activities)}$$

which verifies the basic system of equations:

$$\begin{bmatrix} 4 & 2 \\ 3 & 6 \end{bmatrix} \begin{bmatrix} x_2 = 10/3 \\ x_1 = 4/3 \end{bmatrix} = \begin{bmatrix} 16 \\ 18 \end{bmatrix}$$

Figure 7.6 illustrates the new feasible basis, $FB = [\mathbf{a}_2, \mathbf{a}_1]$. The positive coefficients of the corresponding basic feasible solution, $x_2 = 10/3$ and $x_1 = 4/3$, are the scalars that stretch the respective basis vectors to form a parallelogram whose main diagonal is the constraint vector $\mathbf{b}$. The sequence of figures 7.3, 7.4, and 7.6 provides a convincing illustration that in order to obtain a basic feasible solution it is necessary for the corresponding basis vectors to form a cone encompassing the $\mathbf{b}$ vector. We must now return to step 2.

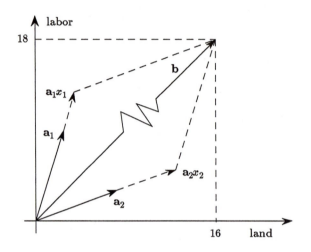

Figure 7.6. Feasible basis and basic feasible solution corresponding to extreme point C.

Iteration #3:

Step 2. There are always two ways to compute the value of the objective function. The first and most demanding in terms of computations evaluates the current basic feasible solution, $BFS = \{x_1 = 4/3,\ x_2 = 10/3,\ x_{s1} = 0,\ x_{s2} = 0\}$, using the original coefficients of the objective function:

$$Z_C = 3\frac{4}{3} + 5\frac{10}{3} + 0(0) + 0(0) = \frac{62}{3}$$

The second procedure builds upon the results previously obtained. The value of the objective function increases from 20 to 62/3 when the basic feasible solution changes from extreme point B to extreme point C:

$$Z_C = Z_B + \frac{1}{2}x_1 = 20 + \frac{1}{2}\left(\frac{4}{3}\right) = \frac{62}{3}$$

The task here is to decide whether the current basic feasible solution is optimal. Since all the productive activities (wheat and corn) are in the feasible basis, it is not convenient (in this small example) to change the product mix by including either unused land or unused labor. The current basic feasible solution is, therefore, optimal.

7.2.4   Recapitulation

In previous sections, the primal simplex algorithm was developed with an eye to the steps to be taken by virtue of their common sense and intuitive economic interpretation. The great appeal of the algorithm relies also upon these aspects as well as the exceptional computational efficiency demonstrated over the years in many different contexts. In this section, however, we wish to show that it is possible to achieve the same results (and more) in a very orderly and easy way by means of a sequence of well-structured tableaux. In fact, it is sufficient to place all the coefficients of the original (P′) problem in an initial primal simplex tableau and apply the pivot method for solving systems of equations as discussed in chapter 6. The only difference is that the choice of the pivot is now dictated by the profitability (entry) and feasibility (exit) criteria developed above.

The sequence of primal simplex tableaux goes from an initial, to an intermediate, to a final tableau. Each tableau except the final one corresponds to an iteration and exhibits the numerical results associated with both steps 2 and 3 of the algorithm. The activity selection based upon the entry criterion is indicated by a vertical arrow, while the activity that leaves the current feasible basis, chosen by means of the exit criterion, is indicated by a horizontal arrow. The coefficient, located at the intersection of the imaginary lines from the two arrows and enclosed by a box, defines the pivot for the current feasible basis. A transformation vector and matrix are defined for each tableau according to the rules developed in chapter 6. The transition from one tableau to the other is performed by carrying out the multiplication between the transformation matrix and the tableau.

The column to the immediate right of each tableau indicates which activities are in the current feasible basis. Because the candidates for this position are several, it is very important to diligently list the indexes of the basic activities. (Basic indexes = BI)

Before plunging into the computations it is useful to recall the two criteria that form the core of the primal simplex algorithm. The entry and exit criteria refer to the primal feasible basis:

> **Entry criterion:** To improve the current feasible solution, select the activity that is associated with the most negative opportunity cost.

The opportunity cost coefficients are located in the row of the objective function.

> **Exit criterion:** To change the current feasible basis, eliminate the column that corresponds to the index in the numerator of the minimum ratio.

The minimum ratio is computed among those ratios with a positive coefficient at the denominator.

Iteration #1: Initial Primal Tableau

$$
T_1 \begin{bmatrix} 1/4 & 0 & \vdots & 0 \\ -3/4 & 1 & \vdots & 0 \\ \cdots \\ 5/4 & 0 & \vdots & 1 \end{bmatrix}
$$

Z	x_1	x_2	x_{s1}	x_{s2}	sol	BI	ratios
0 :	2	4	1	0 :	16	x_{s1}	$16/4 = 4 \rightarrow$ exit
0 :	6	3	0	1 :	18	x_{s2}	$18/3 = 6$
1 :	−3	−5	0	0 :	0	Z_A	

↑
entry

Iteration #2: Intermediate Tableau

$$
T_2 \begin{bmatrix} 1 & -1/9 & \vdots & 0 \\ 0 & 2/9 & \vdots & 0 \\ \cdots \\ 0 & 1/9 & \vdots & 1 \end{bmatrix}
$$

Z	x_1	x_2	x_{s1}	x_{s2}	sol	BI	ratios
0 :	1/2	1	1/4	0 :	4	x_2	$4/(1/2) = 8$
0 :	9/2	0	−3/4	1 :	6	x_{s2}	$6/(9/2)$
							$= 4/3 \rightarrow$ exit
1 :	−1/2	0	5/4	0 :	20	Z_B	

↑
entry

Final (Optimal) Tableau

Z	x_1	x_2	x_{s1}	x_{s2}	P sol	BI
1 :	0	1	1/3	−1/9 :	10/3	x_2
0 :	1	0	−1/6	2/9 :	4/3	x_1
0 :	0	0	7/6	1/9 :	62/3	Z_C
D sol	y_{s1}	y_{s2}	y_1	y_2	R_C	

In the above sequence of tableaux it is possible to identify all the figures and results of the computations developed in the discussion of the primal simplex method. For example, in the intermediate tableau, the opportunity cost of wheat (x_1) is $-1/2$ in the third row, the opportunity labor input requirement of wheat is $9/2$ in the second row, and the marginal rate of technical transformation between corn and wheat is $1/2$ in the first row.

The current basic feasible solution is read in the most right-hand column. Its positive components are identified by the indexes of the basic activities and, therefore, $x_2 = 10/3$ and $x_1 = 4/3$ constitute the optimal solution, as we already know. The nonbasic components of the solution always have a value of zero, $x_{s1} = 0$ and $x_{s2} = 0$. The optimality of the solution is established by the fact that all the opportunity costs in the third row of the final tableau are nonnegative. This implies that any activity outside the current feasible basis is associated with a nonnegative sacrifice and, therefore, it is unprofitable to operate it at positive levels. The optimal value of the objective function is located in the southeast corner of the final tableau and is equal to $62/3 = Z_C$.

The optimal basis is

$$B_{\text{opt}} = \begin{bmatrix} 4 & 2 \\ 3 & 6 \end{bmatrix} = [\mathbf{a}_2 \ \mathbf{a}_1]$$

where the order of the vectors is dictated by the indexes of the basic activities. The inverse of the optimal basis is read from the final tableau and is

$$B_{\text{opt}}^{-1} = \begin{bmatrix} 1/3 & -1/9 \\ -1/6 & 2/9 \end{bmatrix}$$

The inverse is always located under the heading of the slack variables in the place corresponding to the identity matrix I of the initial primal tableau. To verify that B_{opt}^{-1} is indeed the proper inverse, we must check whether $B^{-1}B = I$, according to the rule of chapter 6. Thus,

$$\begin{bmatrix} 1/3 & -1/9 \\ -1/6 & 2/9 \end{bmatrix} \begin{bmatrix} 4 & 2 \\ 3 & 6 \end{bmatrix} \overset{?}{=} \begin{bmatrix} 1 & 0 \\ 0 & 1 \end{bmatrix} : \text{yes}$$

This means that the computations were properly carried out.

7.2.5 Termination of the Primal Simplex Algorithm

The primal simplex algorithm can terminate in any one of the following statuses:

Status #1. A *finite optimal solution* is achieved. This event usually brings bliss to the researcher, but before opening the bottle of champagne it is important to carry out all the numerical checks, especially on the inverse of the optimal basis. A finite optimal solution is detected when both primal basic ($\mathbf{x}_*$) and dual basic ($\mathbf{y}_*$) feasible solutions have been obtained such that $Z_* = \mathbf{c}'\mathbf{x}_* = \mathbf{b}'\mathbf{y}_* = R_*$.

Status #2: An *unbounded primal basic solution* has been found. This implies that the value of the primal objective function can increase without limit. Computationally, this event occurs when it is not possible to find a pivot in the entering column because all the coefficients in that column are either zero or negative.

Status #3. The *infeasibility* of one (or more) constraint is detected. That is, the system of constraints is inconsistent and there is no feasible solution to the given problem. This event is further elaborated in connection with the artificial variable algorithm in chapter 10.

7.3 Dual Variables and the Dual Interpretation of Opportunity Costs

Dual variables were not mentioned in the discussion of the primal simplex algorithm. As we already know, dual variables appear in the dual problem and are interpreted as the imputed prices of land and labor. The dual variables, together with the dual slack variables, constitute the dual solution. Yet it is not necessary to solve the dual problem to find the dual solution. Such a solution is already exhibited in the final tableau. Consider, in fact, the coefficients 7/6 and 1/9 appearing under the headings of the slack variables x_{s1} and x_{s2} in the last row of the final tableau. In section 7.2.3, we argued that opportunity costs are nothing else than transformed price coefficients of the objective function, signifying the profitability of the corresponding commodity relative to the current product mix. Since unused land and unused labor (the meaning of slack variables in our example) are to be interpreted as commodities, we can conclude that those coefficients are the opportunity costs of land and labor. To verify this conclusion, it is sufficient to apply the definition of opportunity cost developed in section 7.2.3, keeping in mind that now both land and labor constraints are binding and the price of the limiting resources in the primal objective function is equal to zero. The allocation of all the available land and labor to productive activities such as wheat and corn calls for a generalization of the notion of marginal rate of technical transformation, as anticipated above. For example, considering the activity of unused land, x_{s1}, its marginal rates of technical transformation in terms of corn and wheat can be determined from the first two rows of the final tableau, which can be read as

$$x_2 + \frac{1}{3}x_{s1} = \frac{10}{3}$$

$$x_1 - \frac{1}{6}x_{s1} = \frac{4}{3}$$

The slack variable for labor, x_{s2}, is kept at zero level. Then, from the first equation, the marginal rate of technical transformation between corn and unused land is

$$MRTT_{x_2,x_{s1}} \equiv -\frac{\partial x_2}{\partial x_{s1}} = \frac{1}{3}$$

From the second equation, the marginal rate of technical transformation between wheat and unused land is

$$MRTT_{x_1,x_{s1}} \equiv -\frac{\partial x_1}{\partial x_{s1}} = -\frac{1}{6}$$

These two coefficients appear in the final tableau under the heading of slack variable x_{s1}. This means that all the coefficients between the vertical dotted lines and above the horizontal dotted line are to be interpreted as marginal rates of technical transformation.

We are ready to verify the meaning of opportunity cost for land and labor. Using the definition stated in section 7.2.3 and recalling that the unit revenue of unused land is zero,

$$\begin{bmatrix} \text{opportunity} \\ \text{cost of land} \end{bmatrix} = \begin{bmatrix} \text{sacrifice in terms} \\ \text{of corn and wheat} \end{bmatrix} - \begin{bmatrix} \text{unit revenue} \\ \text{of land} \end{bmatrix}$$

$$= \begin{bmatrix} \text{unit revenue of} \\ \text{foregone activities} \end{bmatrix} \begin{bmatrix} \text{marginal rates of} \\ \text{technical transformation} \end{bmatrix}$$

$$= \begin{bmatrix} 5 & 3 \end{bmatrix} \begin{bmatrix} 1/3 \\ -1/6 \end{bmatrix} = \frac{5}{3} - \frac{3}{6} = \frac{7}{6}$$

$$\begin{bmatrix} \text{opportunity} \\ \text{cost of labor} \end{bmatrix} = \begin{bmatrix} 5 & 3 \end{bmatrix} \begin{bmatrix} -1/9 \\ 2/9 \end{bmatrix} = -\frac{5}{9} + \frac{6}{9} = \frac{1}{9}$$

As anticipated, the coefficients under the slack variables in the objective function row of the final tableau are also to be interpreted as opportunity costs. Other names are *imputed marginal values, shadow prices,* and *dual variables.* In the vector multiplication of the opportunity cost, the order of revenue coefficients is important and is dictated by the order of the variables in the column of basic indexes.

It remains to verify whether the opportunity costs of land and labor computed above are indeed to be interpreted as dual variables, y_1 and y_2. If true, they must satisfy the dual constraints in the dual problem and give a value of the dual objective function equal to that of the primal.

The dual of the given numerical example is

Dual minimize $R = 16y_1 + 18y_2$

$$\text{subject to} \quad \begin{aligned} 2y_1 + 6y_2 - y_{s1} \qquad &= 3 \\ 4y_1 + 3y_2 \qquad - y_{s2} &= 5 \end{aligned}$$

$$y_1 \geq 0, \ y_2 \geq 0, \ y_{s1} \geq 0, \ y_{s2} \geq 0$$

and the dual solution is a vector of four nonnegative components $\{y_1, y_2, y_{s1}, y_{s2}\}$ that satisfies the dual constraints and minimizes R. Assuming that $y_1 = 7/6$ and $y_2 = 1/9$, $y_{s1} = 0$, $y_{s2} = 0$, the dual constraints are exactly satisfied:

$$\text{constraint \#1}: \ 2\frac{7}{6} + 6\frac{1}{9} = \frac{42 + 12}{18} = \frac{54}{18} = 3$$

$$\text{constraint \#2}: \ 4\frac{7}{6} + 3\frac{1}{9} = \frac{84 + 6}{18} = \frac{90}{18} = 5$$

The value of the dual objective function is equal to that of the primal:

$$\min R = 16\frac{7}{6} + 18\frac{1}{9} = \frac{336 + 36}{18} = \frac{372}{18} = \frac{62}{3} = \max Z$$

The interpretation of dual variables as shadow prices of limiting inputs allows further elaboration of their economic meaning; for example, if we were to increase the land availability by one unit, we would expect that unit of land to be allocated between productive activities corn and wheat in such a way that the additional net revenue would be 7/6 dollars. This notion of marginal changes of inputs evaluated in terms of their contributions to revenue is the familiar value marginal product. Hence, dual variables can also be regarded as value marginal productivities (*VMP*) of the corresponding inputs. Therefore,

$$y_1 = VMP_1 = MR \times MP_1 = \begin{bmatrix} 5 & 3 \end{bmatrix} \begin{bmatrix} 1/3 \\ -1/6 \end{bmatrix} = \frac{7}{6}$$

where $\begin{bmatrix} 5 & 3 \end{bmatrix}$ is the vector of marginal revenues (*MR*) (prices) of the productive activities and

$$\begin{bmatrix} 1/3 \\ -1/6 \end{bmatrix}$$

is the vector of marginal products (*MP*) of land.

One useful element of notation: in order to carry out the discussion in more agile terms and to recognize the notation used in most LP textbooks, we need to restate the notion of opportunity cost in symbolic terms, as seen from a primal perspective. Let

$$
\begin{bmatrix} \text{opportunity} \\ \text{cost of the} \\ j\text{th activity} \end{bmatrix} = \begin{bmatrix} \text{unit revenue} \\ \text{of foregone} \\ \text{activities} \end{bmatrix} \begin{bmatrix} \text{marginal rates} \\ \text{of technical} \\ \text{transformation} \end{bmatrix} - \begin{bmatrix} \text{unit revenue} \\ \text{of the } j\text{th} \\ \text{activity} \end{bmatrix}
$$

$$
= \qquad z_j \qquad - \qquad c_j
$$

where

$$
z_j \equiv \begin{bmatrix} \text{unit revenue} \\ \text{of foregone} \\ \text{activities} \end{bmatrix} \begin{bmatrix} \text{marginal rates} \\ \text{of technical} \\ \text{transformation} \end{bmatrix}
$$

$$
c_j = \begin{bmatrix} \text{unit revenue} \\ \text{of the} \\ j\text{th activity} \end{bmatrix}
$$

The symbol z_j, therefore, is simply a name for foregone income established by pure convention. In subsequent chapters, we will often refer to the opportunity cost of the jth activity simply as $(z_j - c_j)$.

The formulation of opportunity cost elaborated above was stated in terms of *market prices,* since in the firm's framework adopted here, the *unit revenues of foregone activities* represent the market prices of the basic commodities. This version of the opportunity cost can be considered a primal representation.

Naturally, a dual interpretation of opportunity cost defined as *factor costs* exists. This interpretation is obtained directly from the dual constraints. For example, consider the first dual constraint of the numerical illustration, rearranged as follows:

$$
y_{s1} = (2y_1 + 6y_2) - 3
$$

$$
= MC_1 - c_1
$$

$$
= [2 \quad 6] \begin{bmatrix} y_1 \\ y_2 \end{bmatrix} - c_1 = \mathbf{a}_1' \mathbf{y} - c_1 = (z_1 - c_1)
$$

The dual slack variable y_{s1} represents, by definition (section 1.11), the marginal loss of commodity #1. In equivalent words, it is the difference between marginal cost $(2y_1 + 6y_2)$ and marginal revenue c_1. The $\mathbf{a}_1$ vector is the set of input requirements of activity #1, while $\mathbf{y}$ is the vector of input shadow prices. The quantity $a_1'y$ can therefore be interpreted as income foregone, since it is the unit cost of activity #1.

The primal and dual definitions of opportunity cost for the jth activity are summarized as follows:

$$(z_j - c_j) = \begin{cases} \mathbf{mrtt}_j'\mathbf{c}_B - c_j & \text{evaluated at market prices (\textbf{primal})} \\ \mathbf{a}_j'\mathbf{y} - c_j & \text{evaluated at factor costs (\textbf{dual})} \end{cases}$$

where $\mathbf{mrtt}_j$ is the vector of marginal rates of technical transformation of the jth activity and $\mathbf{c}_B$ is the vector of market prices of the basic activities.

The primal and the dual definitions of opportunity cost are equivalent from a numerical viewpoint. To verify this statement, we compute $y_{s1} = (z_1 - c_1)$ using the two definitions and the relevant information contained in the initial and final tableaux:

Primal: $y_{s1} = (z_1 - c_1) = \mathbf{mrtt}_1'\mathbf{c}_B - c_1 = \begin{bmatrix} 0 & 1 \end{bmatrix} \begin{bmatrix} 5 \\ 3 \end{bmatrix} - 3 = 0$

Dual: $y_{s1} = (z_1 - c_1) = \mathbf{a}_1'\mathbf{y} - c_1 = \begin{bmatrix} 2 & 6 \end{bmatrix} \begin{bmatrix} 7/6 \\ 1/9 \end{bmatrix} - 3 = 0$

The computed value for $y_{s1} = (z_1 - c_1)$ is the same one appearing in the last row of the final tableau under the heading of x_1. Therefore, all the $(z_j - c_j)$ coefficients in this last row possess the dual interpretation of opportunity cost. This correspondence suggests stating the final tableau of a linear programming problem in symbolic terms to highlight the meaning of all its components and their symmetric relationships.

Final Tableau
activity levels

Z	primal variables			primal slack variables			P sol
	x_1	$\ldots$	x_n	x_{s1}	$\ldots$	x_{sm}	
0	mrtt_{11}	$\ldots$	mrtt_{1n}	mrtt_{1s1}	$\ldots$	mrtt_{1sm}	x_{B1}
$\ldots$							$\ldots$
0	mrtt_{m1}	$\ldots$	mrtt_{mn}	mrtt_{ms1}	$\ldots$	mrtt_{msm}	x_{Bm}
$\ldots$							$\ldots$
1	$(z_1 - c_1)$	$\ldots$	$(z_n - c_n)$	$(z_{s1} - c_{s1})$	$\ldots$	$(z_{sm} - c_{sm})$	Z_B
D sol	y_{s1}	$\ldots$	y_{sn}	y_1	$\ldots$	y_m	R_B
	dual slack variables			dual variables			

opportunity costs

The terminology "P sol" and "D sol" stands for primal and dual solution, respectively. The above symmetric display is a remarkable achievement of the simplex method. Notice how precisely all the components of the tableau acquire a clear economic interpretation. We recognize, once again, the natural, symmetric duality between primal variables (x_j) and dual slack variables (y_{sj}) and between dual variables (y_i) and primal slack variables (x_{si}).

7.4   Summary of the Primal Simplex Algorithm

Before attempting the implementation of the primal simplex algorithm, we must know an explicit basic feasible solution. The algorithm then is organized around the execution of a two-step procedure. The first step determines the activity that will improve the current basic feasible solution by choosing the most negative opportunity cost. This means selecting the activity for entry in the next feasible basis. The second step determines the activity that must leave the current feasible basis to be replaced by the activity selected in step #1. Symbolically, we can indicate the two steps as follows:

Step #1:

Criterion for entering the primal basis: Select the kth activity such that $(z_k - c_k) = \min_j \{(z_j - c_j) \mid (z_j - c_j) < 0\}$.

Step #2:

Criterion for exiting the primal basis: Select the rth activity such that $\dfrac{x_{Br}}{a_{rk}^*} = \min_i \left\{ \left. \dfrac{x_{Bi}}{a_{ik}^*} \right| a_{ik}^* > 0 \right\}$

The elements a_{ik}^* are the coefficients of the kth activity. In order to maintain the feasibility of the next basis, only positive coefficients are candidates for the denominators of the minimum ratio. The quantities at the numerator are the elements of the current feasible solution, which can be zero. A zero minimum ratio, therefore, is an admissible indicator for the activity that must leave the current feasible basis.

Chapter 8

The Dual Simplex Algorithm

8.1 Introduction

In chapter 7, we studied the primal simplex algorithm, which at that time was regarded as a procedure to find a LP solution working on the primal problem. Analogously, a procedure that solves a dual linear programming problem may be called a dual simplex algorithm. This algorithm was discovered by C. E. Lemke in 1954, seven years after the primal simplex procedure. It is fair to say that without the dual simplex algorithm modern computer codes could not be as reliable as they are. At present, any respectable computer program based on the simplex method incorporates both the primal and the dual simplex algorithms. The reason for the importance of the dual simplex algorithm has its source in the duality and symmetry of linear programming. That it took seven years to discover it is somewhat of a mystery given the obvious symmetry of linear programming.

The original description of the dual algorithm is not easily accessible to the level of this book. Fortunately, however, it is possible to use the notions of duality and symmetry to develop in a rigorous fashion all the rules of the dual procedure. Indeed, the development of the dual simplex algorithm will be done using only our knowledge of the primal simplex algorithm applied to a properly specified dual problem.

It is important to alert the reader to the fact that the application of the primal simplex algorithm to a respecified dual problem concerns only the rules of the algorithm and not the numerical values of the problem. In other words, we are embarking on a metamathematical reasoning where only the form is important and not the content of the problem.

In chapter 7, we mentioned also that it is possible to use the toolbox of a Ferrari to work on a Lamborghini. This fact suggests that the Lamborghini's toolbox will work on a Ferrari. The same relationship of exchangeability exists between LP dual problems and simplex algorithms. The strategy of this chapter is, therefore, as follows: we will use the primal simplex algorithm to work on the dual LP problem and in so doing we will discover the dual simplex algorithm. As for the Lamborghini toolbox that works on a Ferrari, it is possible to apply the dual simplex algorithm to the primal problem because of the discovery of important and novel duality relations.

Let us recall that the development of the primal simplex algorithm was performed using a specification of the primal problem in the form of a maximization subject to inequality constraints of the less-than-or-equal type. The only requirement for using the primal simplex algorithm on the dual problem, then, is to respecify the dual problem in the form of the primal as we have studied it. Consider, therefore, the following dual pair of linear programming problems:

Primal		**Dual**	
maximize $Z =$	$x_1 - 5x_2$	minimize $R =$	$6y_1 + 4y_2$
subject to	$2x_1 - 3x_2 \leq 6$	subject to	$2y_1 - y_2 \geq 1$
	$-x_1 + 2x_2 \leq 4$		$-3y_1 + 2y_2 \geq -5$
	$-x_1 \quad\quad \leq 0$		$y_1 \quad\quad \geq 0$
	$- x_2 \leq 0$		$y_2 \geq 0$

Neither the numerical values of the coefficients nor the size of the problem have any particular significance beyond that of providing a readily manageable LP example. The nonnegativity constraints on the primal variables are written in the form of less-than-or-equal constraints in order to emphasize the symmetric representation of the two problems. We need now to restate the dual problem in the form of the primal. This can be accomplished by multiplying the dual objective function and the dual constraints by (-1):

Dual maximize $(-R) = -6y_1 - 4y_2$

$$\text{subject to} \quad \begin{aligned} -2y_1 + y_2 &\leq -1 \\ 3y_1 - 2y_2 &\leq 5 \\ - y_1 \quad\quad &\leq 0 \\ - y_2 &\leq 0 \end{aligned}$$

To fully understand this second equivalent specification of the dual problem, we recall that, for linear functions, min $R = -$max $(-R)$. Furthermore, the multiplication of an inequality constraint by (-1) changes the direction of the inequality. The dual problem is now written as a maximization problem subject to less-than-or-equal constraints as the primal: the structure (form) of the two problems is the same. This fact allows for the conceptual application of the primal simplex algorithm to the respecified dual problem.

The simplex tableaux associated with the two problems are as follows:

Primal Tableau

Z		x_1	x_2	x_{s1}	x_{s2}		sol
0	:	2	−3	1	0	:	6
0	:	−1	2	0	1	:	4
0	:	−1	0	0	0	:	0
0	:	0	−1	0	0	:	0
...							
1	:	−1	5	0	0	:	0

Dual Tableau

$-R$		y_1	y_2	y_{s1}	y_{s2}		sol
0	:	−2	1	1	0	:	−1
0	:	3	−2	0	1	:	5
0	:	−1	0	0	0	:	0
0	:	0	−1	0	0	:	0
...							
1	:	6	4	0	0	:	0

To proceed in the development of the dual simplex algorithm, it is important to abstract from the specific numerical example. Symbolically, therefore, the two tableaux can be represented as follows:

Primal Tableau

0	:	a_{ij}^*	:	x_{Bi}
...				
1	:	$(z_j - c_j)$	:	Z

Dual Tableau

0	:	$-a_{ji}^*$	:	$(z_j - c_j)$
...				
1	:	x_{Bi}	:	$-R$

The correspondence between the numerical elements of the primal and dual tableaux allows for their symbolic representation as presented above. We know that the coefficients of the primal objective function are interpreted as opportunity costs $(z_j - c_j)$. These coefficients also appear in the dual tableau as the dual solution column. Similarly, the elements of the primal solution x_{Bi} in the primal tableau appear as coefficients of the objective function in the dual tableau. Finally, the technical coefficients a_{ij}^* in the primal tableau appear transposed with a negative sign $(-a_{ji}^*)$ in the dual tableau. The interchange of indexes from (i, j) to (j, i) indicates precisely this transposition. The two tableaux, therefore, contain exactly the same amount of information.

In chapter 7, we discussed the simplex algorithm applied to the primal problem. It consists of two steps: the *entry criterion* and the *exit criterion*, in that order. The two criteria for the *primal simplex algorithm* can be specified more precisely by using the symbolic primal tableau.

The primal simplex algorithm:

Primal Tableau

$$\begin{bmatrix} 0 & \vdots & a_{ij}^* & \vdots & x_{Bi} \\ \multicolumn{5}{c}{\dotfill} \\ 1 & \vdots & (z_j - c_j) & \vdots & Z \end{bmatrix} \rightarrow \quad \min\left\{ \left. \frac{x_{Bi}}{a_{ij}^*} \right| a_{ij}^* > 0 \right\}$$

second step : choose the minimum ratio

exit criterion

first step : choose the most negative $(z_j - c_j)$

entry criterion

The two steps must be interpreted as follows:

First step: To enter the next primal feasible basis, select the activity that is associated with the most negative opportunity cost.

Second step: To exit the current primal feasible basis, select the constraint whose index is associated with the numerator of the minimum ratio criterion.

The minimum ratio criterion expressed in symbolic form must be interpreted as follows: select the minimum ratio between the elements of the current feasible solution (some of which could be equal to zero) and the corresponding (by row) positive elements of the activity selected by the entry criterion. It is crucial to notice the emphasis on the positive elements that go at the denominators. Zero and negative elements should never be considered in the denominator of the minimum ratio to avoid either undefined or infeasible quantities.

8.2 The Dual Simplex Algorithm

With the steps and criteria of the primal simplex algorithm firmly established in relation to the primal tableau, we can proceed to apply the same steps and criteria to the dual tableau, since it corresponds to a problem that has the same form as the primal problem. Notice that we advocate applying the structure of those steps and criteria to the form of the dual tableau and not to its content. This purely logical transposition, based upon the symmetry of linear programming, allows us to discover the criteria of the *dual simplex algorithm* with great simplicity and elegance. It is exactly at this point that the power of a metamathematical reasoning produces surprising results.

The dual simplex algorithm:

Dual Tableau

$$
\begin{bmatrix}
0 & \vdots & -a_{ji}^* & \vdots & (z_j - c_j) \\
\hdotsfor{5} \\
1 & \vdots & x_{Bi} & \vdots & -R
\end{bmatrix}
$$

second step : choose the minimum ratio

$$
\rightarrow \quad \min\left\{\left.\frac{(z_j - c_j)}{-a_{ji}^*}\right| -a_{ji}^* > 0\right\}
$$

exit criterion

first step : choose the most negative x_{Bi}
entry criterion

The two steps relating to the dual tableau must be interpreted as follows:

> **First step:** To enter the next dual feasible basis, select the constraint that is associated with the most negative values of the activity levels.

> **Second step:** To exit the current dual feasible basis, select the activity whose index is associated with the numerator of the minimum ratio criterion.

By comparing the expressions appearing in the primal and dual tableaux, it is possible to deduce very important dual relations. For example, we already know that, for a primal problem, the feasibility of the primal solution requires that all its components x_{Bi} be nonnegative. Furthermore, the optimality of the primal solution requires that all the opportunity costs $(z_j - c_j)$ be nonnegative. These elements appear transposed in the dual tableau. Therefore, we can conclude that, for a dual problem, the feasibility of the dual solution requires that all opportunity costs $(z_j - c_j)$ be nonnegative, while the optimality of the same solution requires that all the activity levels x_{Bi} be nonnegative. It is convenient to restate these conclusions in a more graphic form:

From Primal Tableau		From Dual Tableau
Primal feasibility criterion $x_{Bi} \geq 0$ all i's	$\longleftrightarrow$	Dual optimality criterion $x_{Bi} \geq 0$ all i's
Primal optimality criterion $(z_j - c_j) \geq 0$ all j's	$\longleftrightarrow$	Dual feasibility criterion $(z_j - c_j) \geq 0$ all j's

These conclusions are astonishing. They indicate that the solution of the primal problem signals both feasibility and optimality (or the lack of them) for the primal and dual problems, respectively. Similarly, the solution of the dual problem signals both optimality and feasibility (or the lack of them) for the primal and dual problem, respectively. We thus have:

Primal feasibility is equivalent to **dual optimality** (and the converse)

Primal optimality is equivalent to **dual feasibility** (and the converse)

These symmetric dual relations constitute the essential beauty of the simplex method. It turns out that they are also very useful for solving LP problems. For example, since the information contained in the dual tableau is the same as that of the primal tableau, we can use only one of them to perform the computations of both primal and dual simplex algorithms. The tableau that exhibits a smaller number of constraints should be preferred.

We must pause for a moment to emphasize the remarkable results achieved in the previous pages:

1. We began the chapter with only the knowledge of the primal simplex algorithm as applied to a primal problem stated in the form of a maximization subject to less-than-or-equal constraints.

2. We respecified the associated dual problem into the form of the primal problem and transferred the information of the two problems into the corresponding simplex tableaux. This operation revealed that the two tableaux contained exactly the same information, although arranged in a slightly different fashion: the coefficients of the objective functions and constraints were interchanged and the technological matrix was transposed with a change of sign.

3. The application of the structure of the primal simplex algorithm to the dual tableau expressed in symbolic notation produced a new algorithm called the dual simplex method.

4. From the analysis of the primal and dual tableaux, symmetric pairs of dual relations involving the feasibility and the optimality of the primal and dual solutions were uncovered.

5. Since both tableaux contain the same information, the computations for solving a LP problem will be executed using only the tableau that displays the smaller number of rows. The criteria of both primal and dual simplex algorithms will be applied to that tableau, as required by the status of the primal and dual solutions. In general, the tableau selected for computing the LP solution is the primal tableau. The consequence of applying the dual simplex rules to the primal tableau is a reversal of entry and exit criteria because now the dual algorithm refers to the primal basis.

The steps and criteria of the two simplex algorithms are conveniently compared in table 8.1. Let the subscripts B and NB stand for basic and nonbasic indexes, respectively. Furthermore, PT and DT stand for primal and dual tableau, respectively.

Table 8.1. The Simplex Algorithms

	Primal Simplex	**Dual Simplex**
Initial Condition	A *primal basic feasible* solution must be known: $x_{Bi} \geq 0$, $x_{NBj} = 0$ all i's and j's	A *dual basic feasible* solution must be known: $(z_j - c_j) \geq 0$ all j's
First Step	*Primal entry criterion* (for PT): Select the kth activity such that $(z_k - c_k) = \min_j \{(z_j - c_j)$ given that $(z_j - c_j) < 0\}$	*Dual entry criterion* (for DT): Select the rth activity such that $x_{Br} = \min_i \{x_{Bi} \mid x_{Bi} < 0\}$ *Dual exit criterion* (PT)
Second Step	*Primal exit criterion* (for PT): Select the rth activity such that $\dfrac{x_{Br}}{a^*_{rk}} = \min_i \left\{ \dfrac{x_{Bi}}{a^*_{ik}} \middle\| a^*_{ik} > 0 \right\}$	*Dual exit criterion* (for DT): Select the kth activity such that $\dfrac{(z_k - c_k)}{-a^*_{kr}} = \min_j$ of $\left\{ \dfrac{(z_j - c_j)}{-a^*_{jr}} \middle\| -a^*_{jr} > 0 \right\}$ *Dual entry criterion* (PT)

8.3 Termination of the Dual Simplex Algorithm

As for the primal simplex, the dual simplex algorithm can terminate for the occurrence of one of the following three events:

Status #1. Finite optimal solutions to the given problem have been found. This means that we have found a dual feasible solution $\mathbf{y}_*$ and a primal feasible solution $\mathbf{x}_*$ that produce the equality between the corresponding objective functions $Z_* = \mathbf{c}'\mathbf{x}_* = \mathbf{b}'\mathbf{y}_* = R_*$.

Status #2. An unbounded dual basic solution has been found. The meaning of unboundedness in the dual problem is that the dual objective function can decrease without limit. Computationally, this event occurs when it is not possible to find a pivot in the exit row because all the coefficients are either zero or positive.

Status #3. One (or more) infeasibility(ies) is (are) detected in the system of constraints. The topic of infeasibility is discussed in greater detail in chapter 10.

8.4   An Application of the Dual Simplex Algorithm

The following numerical example provides an interesting illustration of the working of the dual simplex algorithm as well as an introduction to the important topic of multiple optimal solutions. Consider the following primal problem:

Primal maximize $Z = -(3/4)x_1 - x_2$

subject to
$$x_1 + x_2 \geq 1$$
$$2x_1 + 3x_2 \geq 2$$

$$x_1 \geq 0, \; x_2 \geq 0$$

We follow the familiar procedure in setting up the initial primal tableau where all the equality and inequality signs are aligned (although not shown) to the immediate left of the constraint coefficients:

Initial Primal Tableau

$$
\begin{array}{ccccccc}
Z & x_1 & x_2 & x_{s1} & x_{s2} & & \text{sol}
\end{array}
$$

$$
\left[
\begin{array}{cccccc}
0 & : & -1 & -1 & 1 & 0 & : & -1 \\
0 & : & -2 & -3 & 0 & 1 & : & -2 \\
\hdashline
1 & : & 3/4 & 1 & 0 & 0 & : & 0
\end{array}
\right]
$$

Inspection of the initial primal tableau reveals that all the opportunity costs, that is the $(z_j - c_j)$ coefficients, are nonnegative and, therefore, they cannot provide any guidance in choosing an activity for entering the primal basis according to the primal simplex algorithm. Furthermore, the current primal solution is infeasible because it exhibits negative values of the activity levels. This event precisely illustrates the scenario that calls for the application of the dual simplex algorithm. By checking the prescriptions of the algorithm outlined in the previous section, we can verify that, when all

$(z_j - c_j) \geq 0$, we have a dual feasible solution. We must, therefore, begin the application of the dual simplex algorithm with the dual exit criterion from the primal tableau by identifying the most negative component of the primal solution. We will continue with the second step of the algorithm corresponding to the identification of the activity that will enter the primal basis and requires computing the minimum ratio criterion between the elements of the dual feasible solution $(z_j - c_j)$ and the negative coefficients of the row selected in the first step. The intersection of the arrows identifies the pivot. In the following tableaux, the entire application of the dual simplex algorithm is brought to conclusion.

Initial Primal Tableau

$$
T_1
\begin{bmatrix}
1 & -1/3 & \vdots & 0 \\
0 & -1/3 & \vdots & 0 \\
\cdots & \cdots & & \cdots \\
0 & 1/3 & \vdots & 1
\end{bmatrix}
$$

	Z	x_1	x_2	x_{s1}	x_{s2}		sol	BI
	0	$\vdots$ −1	−1	1	0	$\vdots$	−1	x_{s1}
	0	$\vdots$ −2	$\boxed{-3}$	0	1	$\vdots$	−2	x_{s2} →
	1	$\vdots$ 3/4	1	0	0	$\vdots$	0	Z

$$
\text{entry criterion}: \min\left\{\frac{3/4}{-(-2)}, \frac{1}{-(-3)}\right\} = \frac{1}{3}
$$

Intermediate Primal Tableau

$$
T_2
\begin{bmatrix}
-3 & 0 & \vdots & 0 \\
2 & 1 & \vdots & 0 \\
\cdots & \cdots & & \cdots \\
1/4 & 0 & \vdots & 1
\end{bmatrix}
$$

	Z	x_1	x_2	x_{s1}	x_{s2}		sol	BI
	0	$\vdots$ $\boxed{-1/3}$	0	1	−1/3	$\vdots$	−1/3	x_{s1} →
	0	$\vdots$ 2/3	1	0	−1/3	$\vdots$	2/3	x_2
	1	$\vdots$ 1/12	0	0	1/3	$\vdots$	−2/3	Z

$$
\text{entry criterion}: \min\left\{\frac{1/12}{-(-1/3)}, \frac{1/3}{-(-1/3)}\right\} = \frac{1}{4}
$$

Optimal Primal Tableau

Z	x_1	x_2	x_{s1}	x_{s2}		P sol	BI
0	$\vdots$ 1	0	−3	1	$\vdots$	1	x_1
0	$\vdots$ 0	1	2	−1	$\vdots$	0	x_2
1	$\vdots$ 0	0	1/4	1/4	$\vdots$	−3/4	Z

D sol $\quad y_{s1} \quad y_{s2} \quad y_1 \quad y_2$

From the optimal tableau we identify the primal (P sol) and dual (D sol) optimal solutions, the indexes of the activities that constitute the optimal primal basis, its inverse, and the optimal value of the objective function:

optimal primal solution $\{x_1 = 1,\ x_2 = 0,\ x_{s1} = 0,\ x_{s2} = 0\}$

optimal dual solution $\{y_1 = 1/4,\ y_2 = 1/4,\ y_{s1} = 0,\ y_{s2} = 0\}$

optimal primal basis $\begin{bmatrix} -1 & -1 \\ -2 & -3 \end{bmatrix}$ (from the initial tableau)

inverse of the optimal basis $\begin{bmatrix} -3 & 1 \\ 2 & -1 \end{bmatrix}$ (from the final tableau)

optimal value of the objective function $Z_* = -3/4$

To verify the validity of the computations, we first check that the product of the basis and its inverse is equal to the identity matrix:

$$\begin{bmatrix} -1 & -1 \\ -2 & -3 \end{bmatrix} \begin{bmatrix} -3 & 1 \\ 2 & -1 \end{bmatrix} = \begin{bmatrix} 1 & 0 \\ 0 & 1 \end{bmatrix}$$

secondly, that the primal and dual solutions are feasible, and thirdly, that the values of the primal and dual objective functions are equal:

$$Z_* = -3/4 = -3/4(1) - 1(0) = -1(1/4) - 2(1/4) = -3/4 = R_*$$

The optimal primal solution exhibits a basic variable at the zero level, that is, $x_2 = 0$. This event corresponds to a *degenerate primal optimal solution*.

8.5 Degenerate and Multiple Optimal Solutions

A basic solution with one or more basic variables equal to zero is a *degenerate solution*. What does it mean? How does it come about? What are the consequences? To answer these questions, it is useful to graph the primal problem first as in figures 8.1a and 8.1b.

The meaning of degeneracy can be gleaned from figure 8.1a, where the boundary lines of three constraints (constraints #1, #2, and #3) intersect at the same point C. In general, a point is defined by the intersection of two lines. When three lines intersect at the same point, one of them can be considered redundant. From figure 8.1b we deduce that degeneracy means that the constraint vector **b** lies on an edge (a face) of the cone generated by the basic vectors. In this numerical example, the vector **b** is coincident with the vector $\mathbf{a}_1$. In general, we would expect the vector **b** to lie in the interior

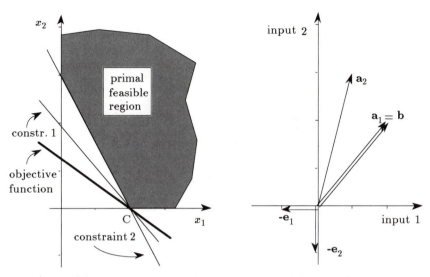

Figure 8.1. (a) Primal output space. (b) Primal input space.

of the basic cone and to represent the main diagonal of the parallelogram whose sides are the basic vectors. When the **b** vector lies on a face of the basic cone, as in the above case, one (or more) basic vector(s) is (are) not needed in the construction of the cone, and the associated variable has a zero value. In this case, the vector a_2 is not needed and $x_2 = 0$.

The consequences of degeneracy in the optimal primal solution are discovered by graphing the dual problem as in figure 8.2. We remind the reader that in order to write down the dual problem correctly, a very helpful step is to specify the primal problem as a maximization subject to less-than-or-equal constraints. With that version of the primal we can use all the conventions set forth in previous chapters. The primal and dual specifications of the above numerical example are as follows:

Primal	**Dual**

maximize $Z = -(3/4)x_1 - x_2$ minimize $R = -y_1 - 2y_2$

subject to
$$-x_1 - x_2 \leq -1$$
$$-2x_1 - 3x_2 \leq -2$$
$$x_1 \geq 0, \; x_2 \geq 0$$

subject to
$$-y_1 - 2y_2 \geq -3/4$$
$$-y_1 - 3y_2 \geq -1$$
$$y_1 \geq 0, \; y_2 \geq 0$$

The dual output space is presented in figure 8.2. The dual objective

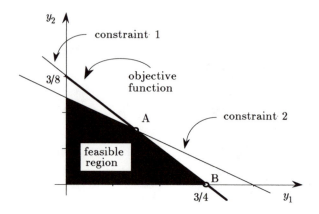

Figure 8.2. Dual output space.

function coincides with a face of the feasible region. Extreme points A and B, therefore, are both optimal basic dual solutions. Furthermore, every point on the A−B segment represents an optimal (nonbasic) dual solution. We conclude that degeneracy in the primal optimal solution is a requirement for multiple optimal solutions in the dual problem. By duality and symmetry, we can extrapolate that degeneracy in the dual solution is a requirement for multiple optimal solutions in the primal problem.

The two optimal solutions of the dual problem corresponding to extreme points A and B can be identified in the simplex tableau of the given example. From the optimal tableau we recognize extreme point A of figure 8.2 as the optimal dual solution:

$$A : \{y_1 = 1/4,\ y_2 = 1/4,\ y_{s1} = 0,\ y_{s2} = 0\}$$

To compute the second optimal dual solution corresponding to extreme point B of figure 8.2, it is necessary to carry out another iteration on the optimal tableau using the dual simplex algorithm. Why the dual simplex algorithm? Because the optimal tableau exhibits all $(z_j - c_j) \geq 0$ (dual feasibility). Furthermore, the same tableau presents a degenerate primal solution due to the presence of a basic variable at zero level, x_2. To trigger the idea of using the dual simplex algorithm, it is convenient to think that the zero value in the primal solution column is associated with a minus $(-)$ sign. These two facts point unequivocally to the use of the dual simplex algorithm.

The computations for the second dual optimal solution are performed in the following tableaux:

First Optimal Tableau

$$
\begin{array}{c}
T_3 \\
\left[\begin{array}{rr:r}
1 & 1 & 0 \\
0 & -1 & 0 \\
\hdashline
0 & 1/4 & 1
\end{array}\right]
\end{array}
\quad
\begin{array}{cccccccc}
Z & x_1 & x_2 & x_{s1} & x_{s2} & & \text{sol} & \text{BI} \\
\left[\begin{array}{c:ccccc:c}
0 & 1 & 0 & -3 & 1 & & 1 \\
0 & 0 & 1 & 2 & \boxed{-1} & & 0 \\
\hdashline
1 & 0 & 0 & 1/4 & 1/4 & & -3/4
\end{array}\right]
& \begin{array}{c} x_1 \\ x_2 \rightarrow \\ \\ Z \end{array}
\end{array}
$$

$$\uparrow$$

entry criterion : minimum ratio

Second Optimal Tableau

$$
\begin{array}{cccccccc}
Z & x_1 & x_2 & x_{s1} & x_{s2} & \text{P sol} & \text{BI} \\
\left[\begin{array}{c:cccc:c}
0 & 1 & 1 & -1 & 0 & & 1 \\
0 & 0 & -1 & -2 & 1 & & 0 \\
\hdashline
1 & 0 & 1/4 & 3/4 & 0 & & -3/4
\end{array}\right]
& \begin{array}{c} x_1 \\ x_{s2} \\ \\ Z \end{array}
\end{array}
$$

D sol $\quad y_{s1} \quad y_{s2} \quad y_1 \quad y_2$

In the second optimal tableau the optimal dual solution corresponds to the extreme point B:

$$B : \{y_1 = 3/4,\ y_2 = 0,\ y_{s1} = 0,\ y_{s2} = 1/4\}$$

The optimal primal solution is essentially the same one appearing in the first optimal tableau even though the basic variable at the zero level has changed. The two optimal basic primal solutions represented by the two optimal tableaux are indistinguishable and, therefore, we say that this numerical example exhibits a unique optimal primal solution and multiple dual optimal solutions.

A final important result: figure 8.2 shows that all the points on the A−B segment of constraint #1 constitute optimal dual solutions. Only the extreme points A and B represent optimal basic solutions, while the infinity of points on the A−B segment are optimal nonbasic solutions. To verify this proposition, it is sufficient to construct one of these nonbasic optimal solutions. First of all, any point on the A−B segment (excluding the end points A and B) is a weighted average of the extreme points A and B with

positive weights that add up to unity. This weighted average is called a
convex combination. Therefore, let the two vectors of basic dual optimal
solutions be given by $\mathbf{y}'_A = (1/4, 1/4, 0, 0)$ and $\mathbf{y}'_B = (3/4, 0, 0, 1/4)$.
Let us also arbitrarily choose weights $\alpha = 1/2$ and $(1-\alpha) = 1/2$. Then the
vector resulting from the convex combination $\mathbf{y} = \alpha \mathbf{y}_A + (1-\alpha)\mathbf{y}_B$ is also
an optimal, but *nonbasic,* dual solution:

$$
\begin{array}{ccccc}
\mathbf{y} & = & \alpha \mathbf{y}_A & + & (1-\alpha)\mathbf{y}_B \\[4pt]
\begin{bmatrix} 1/2 \\ 1/8 \\ 0 \\ 1/8 \end{bmatrix}
& = &
\frac{1}{2}\begin{bmatrix} 1/4 \\ 1/4 \\ 0 \\ 0 \end{bmatrix}
& + &
\frac{1}{2}\begin{bmatrix} 3/4 \\ 0 \\ 0 \\ 1/4 \end{bmatrix}
\end{array}
$$

The dual solution $\mathbf{y}$ is an optimal solution because it satisfies the dual
constraints and produces the same optimal value of the dual objective function associated with extreme points A and B:

$$R_* = b_1 y_1 + b_2 y_2 = -1(1/2) - 2(1/8) = -3/4$$

The $\mathbf{y}$ solution is not a basic solution because it violates the definition of a
basic solution according to which the number of positive elements cannot
be larger than the number of constraints. This is the feature, however, that
makes nonbasic optimal solutions appealing in empirical LP research, since
they offer the possibility of more diversified solutions (production plans) as
compared with basic solutions.

Chapter 9

Linear Programming and the Lagrangean Function

9.1   The Lagrangean Method

In previous chapters we have discussed the relationship between primal and dual problems using the knowledge of economics, as in the theory of the competitive firm. It is important, however, to realize that duality relations are general features of reality and that the principal method for studying them is the *Lagrangean method.*

Proposed around 1790 by Joseph Louis Lagrange, the method has come to be recognized as one of the most successful approaches for studying optimization problems subject to constraints. The theories of the producer and of the consumer could hardly have been analyzed in any meaningful depth without the availability of the Lagrangean method. Briefly stated, the method is based on the knowledge that the maximum of a concave function occurs at a point where the derivative of the function, evaluated at that point, is equal to zero. Lagrange conceived the clever idea of transforming a constrained optimization problem (whose direct solution is usually very difficult) into an unconstrained problem by eliminating the constraints that reappear into a new objective function called the *Lagrangean function.*

As an example, the consumer's problem is defined as that of maximizing a utility function subject to the consumer's available income. More formally, given a utility function $U(q_1, \ldots, q_n)$ and a corresponding budget constraint M, the consumer's problem is stated as

$$\text{maximize} \quad U(q_1, \ldots, q_n)$$

$$\text{subject to} \quad \sum_{i=1}^{n} p_i q_i = M$$

where M is the available income and p_i and q_i are the price and quantity of the ith commodity. The direct solution of this problem is usually impossible because it would require choosing values for the n commodities, verifying whether these values satisfy the constraint, evaluating the utility function and recording its level, and starting the process all over again. The Lagrangean method, on the contrary, makes the solution of the above problem a rational and efficient procedure. The basic idea is to define a Lagrangean function by combining the objective function and the constraint in the following way:

$$L(q_1, \ldots, q_n, \lambda) \equiv U(q_1, \ldots, q_n) + \lambda \left[M - \sum_{i=1}^{n} p_i q_i \right]$$

The function L is now unconstrained and can be optimized by following the familiar rule that sets all its derivatives equal to zero and then solving the resulting system of equations. It is crucial to notice that a new, unknown term λ was introduced in the consumer's problem with the role of penalizing any deviation from the budget constraint. The new variable λ is technically called a *Lagrange multiplier* because it multiplies the constraint, but its real meaning is better revealed by the name of *dual variable*. Dual variables are, therefore, Lagrange multipliers, as confirmed by the following application of the Lagrangean function to a linear programming problem.

In the discussion in chapter 2, the duality relations between a pair of linear programming problems were fully revealed by writing them as follows:

Primal	**Dual**

$$\max Z = c_1 x_1 + c_2 x_2 + c_3 x_3 \qquad\qquad \min R = b_1 y_1 + b_2 y_2$$

subject to $\qquad\qquad\qquad\qquad\qquad\qquad$ subject to

$$
\begin{array}{lcl}
a_{11}x_1 + a_{12}x_2 + a_{13}x_3 \leq b_1 & \longleftrightarrow & y_1 \geq 0 \\
a_{21}x_1 + a_{22}x_2 + a_{23}x_3 \leq b_2 & \longleftrightarrow & y_2 \geq 0 \\
x_1 \geq 0 & \longleftrightarrow & a_{11}y_1 + a_{21}y_2 \geq c_1 \\
x_2 \geq 0 & \longleftrightarrow & a_{12}y_1 + a_{22}y_2 \geq c_2 \\
x_3 \geq 0 & \longleftrightarrow & a_{13}y_1 + a_{23}y_2 \geq c_3
\end{array}
$$

Hence, y_1 and y_2 are dual variables for the primal constraints and x_1, x_2, x_3

are dual variables for the dual constraints. The notion that Lagrange multipliers are in essence dual variables can be illustrated by defining the Lagrangean function for both linear programming problems. The use of the Lagrangean function in this context is of great help in understanding why and how the primal and the dual LP problems are so intimately related to each other.

We now state two Lagrangean functions corresponding to the primal and the dual LP problems only to conclude that the two functions are identical and, therefore, only one Lagrangean function exists for a pair of linear programming problems. In order to maintain the notation adopted for LP, we use y_1, y_2 as Lagrange multipliers for the primal problem and x_1, x_2, x_3 as Lagrange multipliers for the dual. In the definition of the Lagrangean function earlier, the symbol λ was used to indicate the Lagrange multiplier. To reduce the inevitable initial confusion, it is important to realize that λ, x, and y are simply different names (in different contexts) for the same object, the Lagrange multiplier. Therefore, the use of different symbols is legitimate.

Lagrangean for Primal

$$\max L = c_1 x_1 + c_2 x_2 + c_3 x_3$$
$$+ y_1 [b_1 - a_{11} x_1 - a_{12} x_2 - a_{13} x_3]$$
$$+ y_2 [b_2 - a_{21} x_1 - a_{22} x_2 - a_{23} x_3]$$

Lagrangean for Dual

$$\min L^* = b_1 y_1 + b_2 y_2$$
$$+ x_1 [c_1 - a_{11} y_1 - a_{21} y_2]$$
$$+ x_2 [c_2 - a_{12} y_1 - a_{22} y_2]$$
$$+ x_3 [c_3 - a_{13} y_1 - a_{23} y_2]$$

The two Lagrangean functions just stated are actually the same identical function. In this way, the primal and the dual LP problems are tied together by means of a unique Lagrangean function. To show the truth of this proposition, we rewrite the first (primal) Lagrangean by factoring out x_1, x_2, x_3 and rewrite the second (dual) Lagrangean by factoring out y_1, y_2.

Lagrangean for Primal

$$\max L = b_1 y_1 + b_2 y_2$$
$$+ x_1 [c_1 - a_{11} y_1 - a_{21} y_2]$$
$$+ x_2 [c_2 - a_{12} y_1 - a_{22} y_2]$$
$$+ x_3 [c_3 - a_{13} y_1 - a_{23} y_2]$$

Lagrangean for Dual

$$\min L^* = c_1 x_1 + c_2 x_2 + c_3 x_3$$
$$+ y_1 [b_1 - a_{11} x_1 - a_{12} x_2 - a_{13} x_3]$$
$$+ y_2 [b_2 - a_{21} x_1 - a_{22} x_2 - a_{23} x_3]$$

The conclusion is that the two Lagrangean functions are really one, since in each of these functions we can recognize both LP problems.

9.2   The Complementary Slackness Conditions

We recall that the Lagrangean method corresponds to an auxiliary problem set up for the purpose of solving the original (primal or dual) problem. For the success of the procedure, therefore, we must have that, at an optimal point, the value of the Lagrangean function is equal to the value of the objective function in the original problem. That is,

$$\max L = \max Z \qquad\qquad\qquad \min L^* = \min R$$

This implies that each product of a constraint and the corresponding Lagrange multiplier (dual variable) must vanish or, equivalently, that

$$y_1[b_1 - a_{11}x_1 - a_{12}x_2 - a_{13}x_3] = 0 \qquad x_1[c_1 - a_{11}y_1 - a_{21}y_2] = 0$$
$$y_2[b_2 - a_{21}x_1 - a_{22}x_2 - a_{23}x_3] = 0 \qquad x_2[c_2 - a_{12}y_1 - a_{22}y_2] = 0$$
$$x_3[c_3 - a_{13}y_1 - a_{23}y_2] = 0$$

The above zero conditions are called *complementary slackness conditions:* complementary, because the dual variables and the quantity in the brackets complement each other and slackness because the quantities in the brackets represent slack variables.

The complementary slackness conditions possess an important economic interpretation. In order to facilitate the discussion, let us define the following quantities for the first set of complementary slackness conditions:

$$b_1 \quad \equiv \quad \text{availability of resource } \#1$$
$$(a_{11}x_1 + a_{12}x_2 + a_{13}x_3) \quad \equiv \quad \text{use of resource } \#1$$
$$y_1 \quad \equiv \quad \text{VMP of resource } \#1$$

Then the first complementary slackness condition can be analyzed through the following two statements:

$$\text{if } [b_1 - (a_{11}x_1 + a_{12}x_2 + a_{13}x_3)] > 0 \Rightarrow \qquad y_1 \qquad = 0$$
$$\text{if [primal slack variable } \#1] \qquad > 0 \Rightarrow \text{dual variable } \#1 = 0$$

or, in economic terms, when resource #1 is not fully utilized its shadow price is equal to zero; this proposition is directly related to the notion of a free good discussed in chapter 1. The second statement reads:

$$\text{if } y_1 > 0 \qquad\qquad \Rightarrow [b_1 - (a_{11}x_1 + a_{12}x_2 + a_{13}x_3)] = 0$$
$$\text{if dual variable } \#1 > 0 \Rightarrow \quad \text{[primal slack variable } \#1] \quad = 0$$

If the shadow price of resource #1 is positive, it means that the resource is fully utilized in productive activities.

Notice the direction of causality in the above statements. We cannot say that if the primal slack variable is equal to zero then the corresponding dual variable is positive. Complementary slackness conditions as defined above also admit the possibility that both primal slack and dual variables are zero. Although this event has received the name of *dual degeneracy,* it has occurred in empirical work more often than anticipated. The strange name reveals a great deal about the mathematicians' expectations in the early days of linear programming. At that time, it was conjectured that the normal event would have been a combination of either a positive slack and a zero dual variable or of a positive dual and a zero slack variable. Zero slack and dual variables would represent a rare, or degenerate, case. It turned out instead that degeneracy is very common and important, since it is a necessary condition for discovering multiple optimal solutions of LP problems.

The second set of complementary slackness conditions also possesses an important economic interpretation. Let

$$x_1 \equiv \text{output level of activity \#1}$$
$$c_1 \equiv \text{marginal net revenue of activity \#1}$$
$$(a_{11}y_1 + a_{21}y_2) \equiv \text{marginal cost of activity \#1}$$

Two propositions can also be read in this second set of conditions:

$$\text{if } [c_1 - (a_{11}y_1 + a_{21}y_2)] < 0 \quad \Rightarrow \quad x_1 = 0$$
$$\text{if } [\, - \text{ dual slack variable \#1}] < 0 \quad \Rightarrow \quad \text{primal variable \#1} = 0$$

or, in economic terms, if the marginal net revenue of activity #1 is strictly less than its marginal cost, then the optimal strategy is to produce a zero level of commodity #1. The second proposition asserts that

$$\text{if } x_1 > 0 \quad \Rightarrow \quad [c_1 - (a_{11}y_1 + a_{21}y_2)] = 0$$
$$\text{if primal variable \#1} > 0 \quad \Rightarrow \quad [\text{dual slack variable \#1}] = 0$$

if output #1 is produced at positive levels, it means that the marginal cost of activity #1 is equal to its marginal net revenue or, in other words, that activity #1 breaks even.

As for the first set of complementary slackness conditions, in empirical work it is common to encounter the degenerate case when the dual slack and the corresponding primal variables are both zero. This event represents *primal degeneracy.*

The study of the Lagrangean function associated with a pair of linear programming problems has revealed the fundamental role of the primal and dual sets of complementary slackness conditions, which is that of characterizing the primal and dual optimal solutions. The simplex method of linear programming can be interpreted as an efficient numerical procedure for finding values of the primal and dual variables such that they satisfy the complementary slackness conditions.

9.3 Complementary Slackness at Work

The importance of complementary slackness conditions is not limited to the qualitative information discussed in the previous section. Their usefulness can be illustrated in the solution of a linear programming problem without using the primal simplex method. For this purpose, we select a nontrivial numerical example such as

Primal

maximize $Z = 2x_2 + x_3 - 2x_4$

subject to

dual variables

(1)	$x_1 + x_2$	≤ 4	y_1
(2)	$x_1 + 2x_2 + x_3 + x_4$	$= 3$	y_2
(3)	$-2x_1 - x_2 + x_3$	≤ 1	y_3
(4)	x_1	≤ 1	y_4
(5)	x_2	≤ 1	y_5
(6)	x_3	≤ 2	y_6

$$x_j \geq 0, \ j = 1, \ldots, 4$$

All the primal constraints except #2 admit slack variables. Following the familiar rules, the dual problem is specified as follows:

Dual

minimize $R = 4y_1 + 3y_2 + y_3 + y_4 + y_5 + 2y_6$

subject to

primal variables

(a)	$y_1 + y_2 - 2y_3 + y_4$	≥ 0	x_1
(b)	$y_1 + 2y_2 - y_3 + y_5$	≥ 2	x_2
(c)	$y_2 + y_3 + y_6$	≥ 1	x_3
(d)	y_2	≥ -2	x_4

$y_1, \ y_3, \ y_4, \ y_5, \ y_6 \geq 0$, and y_2 free

The second constraint of the primal problem is an equation and, there-fore, the corresponding dual variable is free. All the dual constraints admit slack variables. The dual variables associated with the dual constraints are the primal variables, since the dual of the dual is the primal.

The solution of the above linear programming problem will not use the primal simplex algorithm. Rather, it will be computed using the complementary slackness conditions as a means of finding numerical values of both the primal and dual solutions. When a primal feasible solution and a dual feasible solution have been found, optimal solutions will also be at hand. The procedure must rely upon a sequence of trials.

Trial #1. We begin by selecting the following values for the primal variables:

$$x_1 = x_2 = x_3 = 0 \text{ and } x_4 = 3$$

This selection constitutes a primal feasible solution, since it satisfies all the primal constraints. The corresponding value of the primal objective function is $Z = -6$. Using the information about the primal solution and the complementary slackness rules, we can deduce the following values for the dual variables:

from (1) since $x_{s1} > 0 \Rightarrow y_1 = 0$
from (3) since $x_{s3} > 0 \Rightarrow y_3 = 0$
from (4) since $x_{s4} > 0 \Rightarrow y_4 = 0$
from (5) since $x_{s5} > 0 \Rightarrow y_5 = 0$
from (6) since $x_{s6} > 0 \Rightarrow y_6 = 0$
from (d) since $x_4 > 0 \Rightarrow y_2 = -2$

The dual solution so derived is not feasible, as a check of the dual constraints immediately reveals. For example, from constraint (a) and the fact that nonnegative dual slack variables are subtracted from the left-hand side of the inequality, we have that $(y_2 = -2) \geq 0$, an obvious contradiction. It should be possible to do better in the selection of the values for the primal variables.

Trial #2. Suppose $x_2 = x_3 = 1$ and $x_1 = x_4 = 0$. Then $Z = 3$. This new selection also constitutes a primal feasible solution. The values of the associated dual variables are deduced, as before, through an application of complementary slackness conditions:

from (1) since $x_{s1} > 0 \Rightarrow y_1 = 0$
from (3) since $x_{s3} > 0 \Rightarrow y_3 = 0$
from (4) since $x_{s4} > 0 \Rightarrow y_4 = 0$
from (5) since $x_{s5} > 0 \Rightarrow y_5 \geq 0$
from (6) since $x_{s6} > 0 \Rightarrow y_6 = 0$

but also

from (c) since $x_3 > 0 \qquad\qquad \Rightarrow y_2 = 1$
from (b) since $x_2 > 0$ and $y_2 = 1 \Rightarrow y_5 = 0$

The dual solution obtained in this trial is feasible (check the dual constraints). The value of the dual objective function is $R = 3$, which is equal to the value of the primal objective function. Because we have obtained feasible primal and dual solutions that produce identical objective function values, the linear programming problem is solved (see section 8.2).

Notice the chain of deductions leading to the computation of the dual solution. From constraint (1), with $x_2 = x_3 = 1$, the slack variable $x_{s1} > 0$ and, as a consequence of complementary slackness, $y_1 = 0$. Analogous reasoning is applied to constraints (3), (4), and (6). From constraint (2) it is possible to conclude only that $y_2 \neq 0$. From constraint (5) we can say only that $y_5 \geq 0$ because the corresponding slack variable is equal to zero. The utilization of the information so derived in the dual constraints allows the determination of the values for y_2 and y_5. This example verifies that complementary slackness conditions are necessary and sufficient for finding an optimal solution to a linear programming problem.

Chapter 10

The Artificial Variable Algorithm

10.1 Introduction

With knowledge of the primal simplex and the dual simplex algorithms, it is possible to solve two classes of linear programming problems. The first class contains those problems whose constraints are all inequalities and admit an easy primal feasible basis (the identity matrix I). An equivalent way to regard these problems is to say that the origin of the space (the zero point in the Cartesian coordinate system) belongs to the feasible region. For such problems, a primal basic feasible solution can be established simply by inspection and, therefore, we can find their optimal solution by immediately applying the primal simplex algorithm. The second class contains those problems whose constraints are inequalities and which satisfy the primal optimality criterion (equivalently, the dual feasibility criterion). These problems can be solved by an application of the dual simplex algorithm. It is important to notice that the use of either simplex algorithm requires the availability of an explicit basic feasible solution to either problem. Let us recall that primal feasibility means a nonnegative vector $\mathbf{x}$ that satisfies all the primal constraints. Analogously, dual feasibility means that the dual solution is nonnegative in all its components $(z_j - c_j)$. If this condition is not satisfied, neither simplex algorithm can be applied directly to the LP problem.

Let us suppose, then, that we are confronting a LP problem for which the primal optimality criterion is not satisfied and for which it is difficult to find an initial primal basic feasible solution by inspection. This does not mean that a basic feasible solution for the given problem does not exist, only that it is not easy to find it. We need, therefore, an auxiliary algorithm that will determine whether a primal basic feasible solution exists and, if it exists, the algorithm will find it for us. This procedure is called the *artificial variable algorithm.*

Consider the following LP problem:

Primal (A)

$$\text{maximize } Z = x_1 + 3x_2 + 2x_3 + (1/2)x_4$$

$$\begin{aligned}
\text{subject to} \quad x_1 + 2x_2 + 3x_3 \qquad\qquad\quad &= 15 \\
2x_1 + x_2 + 5x_3 \qquad\qquad\quad &= 20 \\
x_1 + 2x_2 + x_3 + (1/2)x_4 &\le 10
\end{aligned}$$

$$x_j \ge 0 \text{ for all } j\text{'s}$$

We first transform the primal (A) into an equivalent problem with all equality constraints, adding slack variables where possible:

Primal (B)

$$\text{maximize } Z = x_1 + 3x_2 + 2x_3 + (1/2)x_4$$

$$\begin{aligned}
\text{subject to} \quad x_1 + 2x_2 + 3x_3 \qquad\qquad\qquad\quad &= 15 \\
2x_1 + x_2 + 5x_3 \qquad\qquad\qquad\quad &= 20 \\
x_1 + 2x_2 + x_3 + (1/2)x_4 + x_{s3} &= 10
\end{aligned}$$

$$x_j \ge 0 \text{ for all } j\text{'s}, \; x_{s3} \ge 0$$

Clearly, it is difficult to find a basic feasible solution by simple inspection of primal (B) because the first two constraints are equations and, therefore, do not admit slack variables. We deduce, therefore, that the feasible region of this problem does not contain the origin. If a primal basic feasible solution exists, it must involve some three activities out of five. The problem we face is that it is difficult for us to determine by inspection of the primal (B) which three activities constitute a feasible basis.

10.2 Ideas Underlying the Artificial Variable Algorithm

To remedy this situation, the *artificial variable algorithm* comes to the rescue with the following motivation. Suppose we begin with a slightly different problem, artificially constructed, but one that contains the easiest feasible basis, that is, the identity matrix I. In other words, we construct an *artificial problem* whose feasible region contains the origin of the space. A problem of this type is easily formulated by simply introducing one *artificial variable* for each *equality constraint*. In this way, the vectors associated with the artificial variables and the slack vector of the third constraint constitute an artificial feasible basis. That is,

Primal (C)

maximize $Z = x_1 + 3x_2 + 2x_3 + (1/2)x_4$

$$
\begin{array}{rcl}
\text{subject to} \quad x_1 + 2x_2 + 3x_3 + x_{a1} \phantom{+ x_{a2} + x_{s3}} & = & 15 \\
2x_1 + x_2 + 5x_3 \phantom{+ (1/2)x_4 + x_{a1}} + x_{a2} \phantom{+ x_{s3}} & = & 20 \\
x_1 + 2x_2 + x_3 + (1/2)x_4 \phantom{+ x_{a1} + x_{a2}} + x_{s3} & = & 10
\end{array}
$$

$$x_j \geq 0 \text{ for all } j\text{'s}; \ x_{a1} \geq 0, \ x_{a2} \geq 0$$

where x_{a1} and x_{a2} are artificial variables. A basic feasible solution for this expanded artificial problem can now be found by inspection as

$$BFS: \quad \{x_1 = 0, \ x_2 = 0, \ x_3 = 0, \ x_4 = 0, \ x_{a1} = 15, \ x_{a2} = 20, \ x_{s3} = 10\}$$

It should be clear that we are not interested in this artificial problem (C) except as a means to find out whether a basic feasible solution exists for the original linear programming problem (A). Since we have found an explicit basic feasible solution, we can apply the primal simplex algorithm, as we know it, to this artificial problem (C).

If, in the course of the computations carried out strictly following the rules of the primal simplex algorithm, we are able to eliminate the artificial variables from the basic feasible solution (equivalently, we are able to eliminate the artificial vectors from the feasible basis), we must have also found a basic feasible solution for the original problem (A). Conversely, if we cannot eliminate all artificial variables from the basic solution of (C), the original problem (A) is inconsistent, which means that a basic feasible solution for it does not exist.

To guarantee that every effort will be made by the algorithm to eliminate the artificial variables from the basic solution of problem (C), a penalty (cost) will be associated with each artificial variable. In this way, the desire to maximize the objective function will create the conditions such that (if possible) the level of the artificial variables will be reduced to zero, that is, they will be eliminated from the basic solution.

For logical and computational convenience, the penalties associated with the artificial variables are grouped into an *artificial objective function*, W, different from the original objective function, Z. Thus, it appears that we now have two objective functions to deal with. This is indeed the case, but each objective function will guide the selection of activities to enter the feasible basis at different times.

For a maximizing problem such as given in the above example, a suitable artificial objective function is

$$\text{maximize } W = -x_{a1} - x_{a2}$$

The maximum value of W is zero. Because $x_{a1} \geq 0$ and $x_{a2} \geq 0$, the maximum is attained only when $x_{a1} = x_{a2} = 0$. Hence, if in the process of applying the primal simplex algorithm we are able to drive both artificial variables to zero (equivalently, drive the artificial vectors out of the feasible basis), we are assured that $W = 0$. At this stage, we have obtained a basic feasible solution for the original problem and we can continue the computations by applying the primal simplex algorithm to maximize the original objective function Z. (In problems that possess degenerate solutions, however, it is possible to reach a point where $W = 0$, but some artificial vector is still in the feasible basis with the corresponding artificial variable at a zero level in the basic feasible solution. The resolution of this problem will be discussed in section 10.3.)

In summary, the process of obtaining an optimal solution for problem (A) can be divided into two phases.

Phase I deals with problem (C) and attempts to eliminate x_{a1}, x_{a2} from the basic feasible solution. To do so, the entry criterion for selecting the activity to enter the primal basis is based upon the artificial objective function W. When all artificial activity vectors have been driven out of the primal feasible basis, phase I ends with $W = 0$ and with an explicit basic feasible solution for problem (A). If the primal simplex algorithm stops with $W \neq 0$, problem (A) does not have a basic feasible solution. The only recourse is to reformulate the original problem.

Phase II begins with a successful ending of phase I. Now the application of the primal simplex algorithm attempts to maintain the feasibility of the basic solution and optimize the original objective function Z. During phase II, the entry criterion selects activities with reference to opportunity costs in the Z row.

We are now ready to solve the given LP problem following the ideas outlined above. We begin with the setup of the initial primal simplex tableau corresponding to problem (C) plus the artificial objective function W. BI stands for basic indexes.

Initial Primal Tableau

Z	W		x_1	x_2	x_3	x_4	x_{a1}	x_{a2}	x_{s3}		sol	BI
0	0	:	1	2	3	0	1	0	0	:	15	x_{a1}
0	0	:	2	1	5	0	0	1	0	:	20	x_{a2}
0	0	:	1	2	1	1/2	0	0	1	:	10	x_{s3}
0	1	:	0	0	0	0	[1	1]	0	:	0	W
1	0	:	−1	−3	−2	−1/2	0	0	0	:	0	Z

The computations of the primal simplex algorithm cannot begin because the above tableau does not exhibit the easiest feasible basis we are looking for, namely a basis made up of unit vectors, the identity basis I. To achieve that goal we must eliminate the bracketed unit coefficients in the row of the artificial objective function W. This task is accomplished by exploiting a property of linear systems that allows us to subtract, add, and multiply by a scalar entire rows of the system without changing its structural meaning. Hence, the rule for eliminating the bracketed unit values is

> *Add* by columns all the coefficients in the rows containing artificial variables and *subtract* the result from the corresponding coefficients of the artificial objective function.

Initial Primal Tableau

Z	W		x_1	x_2	x_3	x_4	x_{a1}	x_{a2}	x_{s3}		sol	BI
0	0	:	1	2	3	0	1	0	0	:	15	x_{a1}
0	0	:	2	1	5	0	0	1	0	:	20	x_{a2}
0	0	:	1	2	1	1/2	0	0	1	:	10	x_{s3}
0	1	:	−3	−3	−8	0	0	0	0	:	−35	W
1	0	:	−1	−3	−2	−1/2	0	0	0	:	0	Z

Only rows #1 and #2 from the previous tableau have been added together and subtracted from the W row because only these rows contain artificial variables. Phase I can now begin using the artificial objective function W to apply the entry criterion. The rules are those of the primal simplex algorithm. Arrows indicate the selection of the entry and exit criteria. A boxed coefficient at the intersection of the arrows' lines is the pivot.

Phase I begins:

Initial Primal Tableau

$$
\begin{bmatrix}
1 & -3/5 & 0 & \vdots & 0 & 0 \\
0 & 1/5 & 0 & \vdots & 0 & 0 \\
0 & -1/5 & 1 & \vdots & 0 & 0 \\
\cdots\cdots\cdots\cdots\cdots \\
0 & 8/5 & 0 & \vdots & 1 & 0 \\
0 & 2/5 & 0 & \vdots & 0 & 1
\end{bmatrix}
$$

Z	W	x_1	x_2	x_3	x_4	x_{a1}	x_{a2}	x_{s3}	sol	BI	ratios
0	0:	1	2	3	0	1	0	0:	15	x_{a1}	5
0	0:	2	1	$\boxed{5}$	0	0	1	0:	20	x_{a2}	4 →
0	0:	1	2	1	1/2	0	0	1:	10	x_{s3}	10
$\cdots$											
0	1:	−3	−3	−8 ↑	0	0	0	0:	−35	W	
1	0:	−1	−3	−2	−1/2	0	0	0:	0	Z	

Intermediate Primal Tableau

$$
\begin{bmatrix}
5/7 & 0 & 0 & \vdots & 0 & 0 \\
-1/7 & 1 & 0 & \vdots & 0 & 0 \\
-9/7 & 0 & 1 & \vdots & 0 & 0 \\
\cdots\cdots\cdots\cdots\cdots \\
1 & 0 & 0 & \vdots & 1 & 0 \\
13/7 & 0 & 0 & \vdots & 0 & 1
\end{bmatrix}
$$

Z	W	x_1	x_2	x_3	x_4	x_{a1}	x_{a2}	x_{s3}	sol	BI
0	0:	−1/5	$\boxed{7/5}$	0	0	1	−3/5	0:	3	x_{a1} →
0	0:	2/5	1/5	1	0	0	1/5	0:	4	x_3
0	0:	3/5	9/5	0	1/2	0	−1/5	1:	6	x_{s3}
$\cdots$										
0	1:	1/5	−7/5 ↑	0	0	0	8/5	0:	−3	W
1	0:	−1/5	−13/5	0	−1/2	0	2/5	0:	8	Z

Intermediate Primal Tableau

Z	W	x_1	x_2	x_3	x_4	x_{a1}	x_{a2}	x_{s3}	sol	BI
0	0 :	−1/7	1	0	0	5/7	−3/7	0 :	15/7	x_2
0	0 :	3/7	0	1	0	−1/7	2/7	0 :	25/7	x_3
0	0 :	6/7	0	0	1/2	−9/7	4/7	1 :	15/7	x_{s3}
$\cdots$										
0	1 :	0	0	0	0	1	1	0 :	0	W
1	0 :	−4/7	0	0	−1/2	13/7	−5/7	0 :	95/7	Z

End of phase I.

Phase I is successfully terminated. We have achieved a basic feasible solution for the original problem (A):

$$BFS = \{x_2 = 15/7,\ x_3 = 25/7,\ x_{s3} = 15/7,\ \text{and}\ x_1 = x_4 = 0\}$$

All artificial activity vectors were driven out of the feasible basis and, therefore, all artificial variables disappeared from the basic feasible solution; as a consequence, $W = 0$. We no longer need the artificial objective function W, which can be eliminated from the next tableaux. Actually, the artificial activities have also performed their useful role and will no longer be used as candidates for entering the feasible basis. They are maintained in the next tableaux, however, in order to be able to read off all the dual variables and the inverse of the optimal basis. We now continue the computations using the primal simplex algorithm.

Phase II begins:

Intermediate Primal Tableau

		Z	x_1	x_2	x_3	x_4	x_{a1}	x_{a2}	x_{s3}	sol	BI
1	0 1/6 : 0	0:	$-1/7$	1	0	0	$5/7$	$-3/7$	0:	$15/7$	x_2
0	1 $-1/2$: 0	0:	$3/7$	0	1	0	$-1/7$	$2/7$	0:	$25/7$	x_3
0	0 7/6 : 0	0:	$\boxed{6/7}$	0	0	$1/2$	$-9/7$	$4/7$	1:	$15/7$	$x_{s3} \rightarrow$
0	0 4/6 : 1	1:	$-4/7$	0	0	$-1/2$	$13/7$	$-5/7$	0:	$95/7$	Z

$\uparrow$

The entry criterion indicates that $(z_1 - c_1) = -4/7$ is the most negative opportunity cost. The coefficient $(z_{a2} - c_{a2}) = -5/7$, although smaller than $-4/7$, cannot be considered as a selection for the entry criterion because it is associated with an artificial activity that never should reappear in the feasible basis.

Intermediate Primal Tableau

		Z	x_1	x_2	x_3	x_4	x_{a1}	x_{a2}	x_{s3}	sol	BI
1	0 $-1/7$: 0	0:	0	1	0	$1/12$	$1/2$	$-1/3$	$1/6$:	$5/2$	x_2
0	1 3/7 : 0	0:	0	0	1	$-1/4$	$1/2$	0	$-1/2$:	$5/2$	x_3
0	0 12/7 : 0	0:	1	0	0	$\boxed{7/12}$	$-3/2$	$2/3$	$7/6$:	$5/2$	$x_1 \rightarrow$
0	0 2/7 : 1	1:	0	0	0	$-1/6$	1	$-1/3$	$4/6$:	15	Z

$\uparrow$

The selection of the entry activity bypasses the $(-1/3)$ coefficient for the same reason expounded in the previous tableau. Notice that it is perfectly legitimate for an activity to enter the feasible basis and immediately leave it, such as activity #1.

Optimal Primal Tableau

	Z	x_1	x_2	x_3	x_4	x_{a1}	x_{a2}	x_{s3}	P sol	BI
	0	$-1/7$	1	0	0	$5/7$	$-3/7$	0	$15/7$	x_2
	0	$3/7$	0	1	0	$-1/7$	$2/7$	0	$25/7$	x_3
	0	$12/7$	0	0	1	$-18/7$	$8/7$	2	$30/7$	x_4
	1	$2/7$	0	0	0	$4/7$	$-1/7$	1	$110/7$	Z
D sol		y_{s1}	y_{s2}	y_{s3}	y_{s4}	y_1	y_2	y_3		

End of phase II.

The final tableau of phase II is optimal even though it displays a negative opportunity cost such as $y_2 = -1/7$. From a previous discussion we know that in phase II artificial activities should not be considered for entering the feasible basis. Furthermore, the dual variable y_2 is associated with a constraint expressed as an equation, and from chapter 2 we know that such a dual variable is unrestricted. Therefore, a negative value for y_2 is admissible in an optimal solution.

Before leaving this numerical example it is crucial to conduct several tests to verify that indeed the final tableau is also optimal. First of all, the information contained in the final optimal tableau must be put in evidence:

primal solution $= \{x_1 = 0, \; x_2 = 15/7, \; x_3 = 25/7, \; x_4 = 30/7, x_{s3} = 0\}$

dual solution $= \{y_{s1} = 2/7, \; y_{s2} = 0, \; y_{s3} = 0, \; y_{s4} = 0, \; y_1 = 4/7,$
$y_2 = -1/7, \; y_3 = 1\}$

primal basis $= B = [\mathbf{a}_2 \; \mathbf{a}_3 \; \mathbf{a}_4] = \begin{bmatrix} 2 & 3 & 0 \\ 1 & 5 & 0 \\ 2 & 1 & 1/2 \end{bmatrix}$

The information for assembling the optimal primal basis is taken from the final and the initial tableaux. The indexes of the activities that constitute the optimal basis are those listed on the BI (basic indexes) column of the final tableau. The coefficients of those basic activities are taken from either the initial tableau or the original problem. The order of the activities in the optimal basis is dictated by the order in the column of basic indexes:

$$\text{inverse of the optimal primal basis } = B^{-1} = \begin{bmatrix} 5/7 & -3/7 & 0 \\ -1/7 & 2/7 & 0 \\ -18/7 & 8/7 & 2 \end{bmatrix}$$

The inverse of the optimal basis always appears in the portion of the final tableau that corresponds to the identity matrix in the initial tableau. Three basic checks of the computations can be performed using this inverse. A first check verifies that $BB^{-1} = I$; that is, the product of the basis and its inverse must be equal to the identity matrix:

$$\begin{bmatrix} 2 & 3 & 0 \\ 1 & 5 & 0 \\ 2 & 1 & 1/2 \end{bmatrix} \begin{bmatrix} 5/7 & -3/7 & 0 \\ -1/7 & 2/7 & 0 \\ -18/7 & 8/7 & 2 \end{bmatrix} \overset{?}{=} \begin{bmatrix} 1 & 0 & 0 \\ 0 & 1 & 0 \\ 0 & 0 & 1 \end{bmatrix} : \text{yes}$$

A second check verifies that the primal optimal solution was computed correctly by recalculating it as $B^{-1}b = x_B$, where b is the constraint vector of the primal problem and x_B is the nonzero portion of the optimal primal solution:

$$\begin{bmatrix} 5/7 & -3/7 & 0 \\ -1/7 & 2/7 & 0 \\ -18/7 & 8/7 & 2 \end{bmatrix} \begin{bmatrix} 15 \\ 20 \\ 10 \end{bmatrix} \overset{?}{=} \begin{bmatrix} 15/7 \\ 25/7 \\ 30/7 \end{bmatrix} : \text{yes}$$

A third check involves recomputing the dual variables using the coefficients of the primal objective function that are associated with the basic activities to verify that $c_B' B^{-1} = y'$, where c_B' is a row vector of net revenue coefficients:

$$[3 \quad 2 \quad 1/2] \begin{bmatrix} 5/7 & -3/7 & 0 \\ -1/7 & 2/7 & 0 \\ -18/7 & 8/7 & 2 \end{bmatrix} \overset{?}{=} [4/7 \quad -1/7 \quad 1] : \text{yes}$$

A final, useful check consists in reproducing the opportunity cost for any commodity of the final tableau. Consider x_1, whose activity vector was not selected for the optimal basis. Its opportunity cost is displayed at $2/7$ in the last row of the final tableau. The dual computations of this quantity are as follows:

$$y_{s1} = (z_1 - c_1) = \begin{cases} \mathbf{mrtt}_1' c_B - c_1 = \begin{bmatrix} \dfrac{-1}{7} & \dfrac{3}{7} & \dfrac{12}{7} \end{bmatrix} \begin{bmatrix} 3 \\ 2 \\ 1/2 \end{bmatrix} - 1 = \dfrac{2}{7} \\[3em] a_1' y - c_1 \quad = [1 \quad 2 \quad 1] \begin{bmatrix} 4/7 \\ -1/7 \\ 1 \end{bmatrix} - 1 = \dfrac{2}{7} \end{cases}$$

The task of identifying the optimal dual basis remains. The information about the indexes of the dual activities that enter the dual basis is extracted from the last row of the optimal tableau. For nondegenerate dual problems the rule is as follows:

> The indexes of the *nonzero dual variables* (including slack variables) in the final tableau indicate the dual activities that form the optimal dual basis.

For degenerate dual solutions, the identification of the dual activities that constitute the dual basis must rely on the indexes of the nonbasic primal activities. The verification of this rule is done rather simply by first writing down the dual problem and then checking to see that the nonzero dual variables of the final primal tableau constitute a basic feasible solution:

Dual

minimize $R = 15y_1 + 20y_2 + 10y_3$

$$
\begin{array}{rcl}
\text{subject to} \quad y_1 + 2y_2 + y_3 - y_{s1} & = & 1 \\
2y_1 + y_2 + 2y_3 \qquad - y_{s2} & = & 3 \\
3y_1 + 5y_2 + y_3 \qquad\quad - y_{s3} & = & 2 \\
(1/2)y_3 \qquad\qquad\qquad - y_{s4} & = & 1/2
\end{array}
$$

y_1 and y_2 free, $y_3 \geq 0$, $y_{sj} \geq 0$, $j = 1, \ldots, 4$

The last row of the final tableau indicates that $y_{s2} = y_{s3} = y_{s4} = 0$, while the other dual variables are nonzero and satisfy the dual constraints. The dual system of constraints, therefore, reduces to a basic system of four equations in four variables and the associated dual basis is

$$
\text{optimal dual basis} = \begin{bmatrix} 1 & 2 & 1 & -1 \\ 2 & 1 & 2 & 0 \\ 3 & 5 & 1 & 0 \\ 0 & 0 & 1/2 & 0 \end{bmatrix}
$$

A further dual relationship is displayed by the dual basis, which contains the transposed primal basis in the lower southwest part of the matrix. The initial and final primal tableaux, therefore, contain all the information necessary for a complete analysis and interpretation of the LP problem.

When using a commercially available computer program for solving a large LP problem, the inverse of the optimal basis is generally not available to the user. The computer program simply does not print it out. In this case, it is very important to perform checks about the feasibility of the primal and dual solutions that require verifiying as to whether they satisfy the primal and dual constraints, respectively. A further possible check consists in verifying whether the recomputed value of the primal objective function is equal to its dual counterpart.

10.3   Termination of the Artificial Variable Algorithm

The rules of the artificial variable algorithm can produce the following outcomes:

Status #1.   An optimal value of the artificial objective function ($W = 0$) has been achieved in phase I, together with the elimination of all artificial activity vectors from the feasible basis. This event implies that a basic feasible solution for the original problem has been found and we can now proceed to phase II with the application of the primal simplex algorithm.

Status #2.   An unbounded basic feasible solution has been found in phase I. This implies that the original problem is also unbounded.

Status #3.   The rules of the artificial variable algorithm can no longer be applied because all the $(z_j - c_j) \geq 0$ in the W row and $W \neq 0$. This event implies that the original LP problem does not have a basic feasible solution; in other words, the system of constraints is inconsistent.

Sometimes it may happen that we encounter the event $W = 0$ with some artificial activity vector still in the basis and its corresponding variable at the zero level. This is a case of a degenerate artificial basic feasible solution. We should try to remove this artificial activity by replacing it with some other structural activity. If this is possible, we are in status #1. If it is not possible (because we cannot find a pivot), we are in status #3.

10.4   Another Application of the Artificial Variable Algorithm

The artificial variable algorithm can be used when the set of primal constraints contains equations (as in the above numerical example) and, in general, when it is difficult to find a basic feasible solution. Consider this second LP example:

Primal   maximize $Z = 35x_1 + 42x_2$

subject to
$$3x_1 + 2x_2 \geq 10$$
$$6x_1 + x_2 \leq 48$$

$$x_1 \geq 0, \ x_2 \geq 0$$

Adding slack variables we obtain:

Primal   maximize $Z = 35x_1 + 42x_2$

subject to
$$3x_1 + 2x_2 - x_{s1} \qquad = 10$$
$$6x_1 + x_2 \qquad + x_{s2} = 48$$

$$x_1 \geq 0, \ x_2 \geq 0, \ x_{s1} \geq 0 \ x_{s2} \geq 0$$

Still we cannot find a basic feasible solution easily. In fact, if we set $x_1 = x_2 = 0$, then $x_{s1} = -10$ and $x_{s2} = 48$, clearly not a primal feasible solution. The use of the dual simplex algorithm is also prevented because not all the opportunity costs $(z_j - c_j)$ are nonnegative. This type of problem, therefore, calls for an application of the artificial variable algorithm. The above numerical example requires one artificial variable x_{a1} and an artificial objective function defined as: maximize $W = -x_{a1}$.

Initial Primal Tableau

Z	W		x_1	x_2	x_{s1}	x_{a1}	x_{s2}		sol
0	0	:	3	2	−1	1	0	:	10
0	0	:	6	1	0	0	1	:	48
0	1	:	0	0	0	[1]	0	:	0
1	0	:	−35	−42	0	0	0	:	0

As in a previous example, this initial tableau must be transformed to exhibit an explicit identity basis by subtracting the row of coefficients that contains the artificial variable from the row of the artificial objective function. Only then is it possible to read an explicit basic feasible solution and the computations of phase I can begin.

Phase I begins:

Initial Primal Tableau

$$
\begin{bmatrix}
1/3 & 0 & \vdots & 0 & 0 \\
-2 & 1 & \vdots & 0 & 0 \\
\cdots \\
1 & 0 & \vdots & 1 & 0 \\
35/3 & 0 & \vdots & 0 & 1
\end{bmatrix}
$$

	Z	W	x_1	x_2	x_{s1}	x_{a1}	x_{s2}		sol	BI
	0	0	$\vdots$	$\boxed{3}$	2	−1	1	0	$\vdots$ 10	x_{a1} →
	0	0	$\vdots$	6	1	0	0	1	$\vdots$ 48	x_{s2}
			$\cdots$							
	0	1	$\vdots$	−3 ↑	−2	1	0	0	$\vdots$ −10	W
	1	0	$\vdots$	−35	−42	0	0	0	$\vdots$ 0	Z

Intermediate Primal Tableau

Z	W	x_1	x_2	x_{s1}	x_{a1}	x_{s2}		sol	BI
0	0	$\vdots$ 1	2/3	−1/3	1/3	0	$\vdots$	10/3	x_1
0	0	$\vdots$ 0	−3	2	−2	1	$\vdots$	28	x_{s2}
0	1	$\vdots$ 0	0	0	1	0	$\vdots$	0	W
1	0	$\vdots$ 0	−56/3	−35/3	35/3	0	$\vdots$	350/3	Z

End of Phase I.

The row of the artificial objective function is no longer necessary and it is eliminated from the primal tableau. Also the column associated with the artificial activity can be dropped because, in this example, there is a legitimate slack activity $e_1(x_{s1})$ that will be associated with the corresponding dual variable.

Phase II begins:

Intermediate Primal Tableau

$$
\begin{bmatrix}
3/2 & 0 & \vdots & 0 \\
9/2 & 1 & \vdots & 0 \\
\cdots \\
28 & 0 & \vdots & 1
\end{bmatrix}
$$

	Z	x_1	x_2	x_{s1}	x_{s2}		sol	BI
	0	$\vdots$ 1	$\boxed{2/3}$	−1/3	0	$\vdots$	10/3	x_1 →
	0	$\vdots$ 0	−3	2	1	$\vdots$	28	x_{s2}
	1	$\vdots$ 0	−56/3	−35/3	0	$\vdots$	350/3	Z

↑

Intermediate Primal Tableau

$$
\begin{bmatrix}
1 & 1 & \vdots & 0 \\
0 & 2 & \vdots & 0 \\
\cdots \\
0 & 42 & \vdots & 1
\end{bmatrix}
$$

	Z		x_1	x_2	x_{s1}	x_{s2}		sol	BI
	0	$\vdots$	3/2	1	$-1/2$	0	$\vdots$	5	x_2
	0	$\vdots$	9/2	0	$\boxed{1/2}$	1	$\vdots$	43	x_{s2} $\rightarrow$
	1	$\vdots$	28	0	-21	0	$\vdots$	210	Z

$\uparrow$

Optimal Primal Tableau

	Z		x_1	x_2	x_{s1}	x_{s2}	P sol		BI
	0	$\vdots$	6	1	0	1	$\vdots$	48	x_2
	0	$\vdots$	9	0	1	2	$\vdots$	86	x_{s1}
	1	$\vdots$	217	0	0	42	$\vdots$	2016	Z

D sol $\quad y_{s1} \quad y_{s2} \quad y_1 \quad y_2$

End of Phase II. Let us identify the various components of the linear programming problem:

optimal primal solution $\quad = \quad \{x_1 = 0,\ x_2 = 48,\ x_{s1} = 86,\ x_{s2} = 0\}$

optimal dual solution $\quad = \quad \{y_{s1} = 217,\ y_{s2} = 0,\ y_1 = 0,\ y_2 = 42\}$

optimal primal basis $\quad = \quad \begin{bmatrix} -2 & 1 \\ 1 & 0 \end{bmatrix}$

optimal dual basis $\quad = \quad \begin{bmatrix} 6 & -1 \\ 1 & 0 \end{bmatrix}$

A fundamental check of the computations involves the primal basis and its inverse. Hence,

$$
\begin{bmatrix} -2 & 1 \\ 1 & 0 \end{bmatrix}
\begin{bmatrix} 0 & 1 \\ 1 & 2 \end{bmatrix}
=
\begin{bmatrix} 1 & 0 \\ 0 & 1 \end{bmatrix}
$$

Some attention must be given to defining the optimal dual bases for this example. The general rule is that the primal problem should be written in the form of a maximization with *all* constraints exhibiting inequalities in the same direction. When this formulation is achieved, the identification of the optimal bases is a simple task. The dual basis must obviously be identified using the proper dual formulation of the problem.

Chapter 11

The Artificial Constraint Algorithm

11.1 Ideas Underlying the Artificial Constraint Algorithm

When it is difficult to find an initial basic feasible solution for either the primal or the dual problem, the *artificial constraint algorithm* provides a *dual alternative* to the *artificial variable method*. This is so because the dual of a variable is a constraint and vice versa (see chapter 2). Armed with this dual notion (variable $\longleftrightarrow$ constraint), we can infer the requirements of the artificial constraint algorithm from the requirements of the artificial variable algorithm by simply exchanging every dual notion, as in the scheme shown in table 11.1.

The rules of the artificial constraint algorithm will be illustrated by discussing the following LP example:

Primal maximize $Z = \quad 3x_1 + 2x_2 + 4x_3$

$$\text{subject to} \qquad \begin{aligned} x_1 - x_2 + 3x_3 &\leq -1 \\ -x_1 + 2x_2 + x_3 &\leq 3 \end{aligned}$$

$$x_j \geq 0, \ j = 1, \ldots, 3$$

Clearly, no primal basic feasible solution is easily discernible by inspection. In fact, if we let $x_1 = x_2 = x_3 = 0$, then $x_{s1} = -1$ and $x_{s2} = 3$, an infeasible result. Let us, therefore, add the following artificial constraint to the above problem:

$$x_1 + x_2 + x_3 + x_{s0} = M$$

where M is a sufficiently large positive number to avoid being an effective

Table 11.1. The Artificial Algorithms

	Artificial Variable		Artificial Constraint
(a)	A primal basic feasible solution is not easily available.	(a)	A dual basic feasible solution is not easily available.
(b)	Introduce as many artificial variables as necessary to achieve an (artificial) primal basic feasible solution. This operation will make the dual solution a function of the artificial variables.	(b)	Introduce one artificial constraint to achieve an (artificial) dual basic feasible solution. This operation will make the primal solution a function of the artificial constraint.
(c)	Make the problem primal optimal by attempting to free the dual solution from the dependency on the artificial variables. The rules of the primal simplex algorithm are followed during this attempt.	(c)	Make the problem primal feasible by attempting to free the primal solution from the dependency on the artificial constraint. The rules of the dual simplex algorithm are followed during this attempt.

constraint for $(x_1 + x_2 + x_3)$. This requirement is always met rather easily. The new variable x_{s0} represents the slack variable associated with the artificial constraint M.

The initial primal simplex tableau for this enlarged problem is as follows:

Initial Primal Tableau

$$
\begin{array}{c}
 \\
x_{s1} \\
x_{s2} \\
x_{s0} \\
 \\
Z
\end{array}
\quad
\begin{array}{c}
Z \quad x_1 \quad x_2 \quad x_3 \quad x_{s1} \quad x_{s2} \quad x_{s0} \quad \text{sol} \quad \text{BI} \\
\left[
\begin{array}{c:cccccc:cc}
0 & 1 & -1 & 3 & 1 & 0 & 0 & -1 \\
0 & -1 & 2 & 1 & 0 & 1 & 0 & 3 \\
0 & 1 & 1 & 1 & 0 & 0 & 1 & M \\
\hdashline
1 & -3 & -2 & -4 & 0 & 0 & 0 & 0
\end{array}
\right]
\end{array}
$$

In order to implement point (b) of the artificial constraint algorithm stated in table 11.1, that is, in order to make the dual solution feasible, we pivot immediately on a unit coefficient of the artificial constraint corresponding to the most negative $(z_j - c_j)$. When this is done, the next tableau exhibits the following characteristics:

1. A dual basic feasible solution has been established (equivalently, the primal optimality criterion is satisfied).

2. The artificial variable x_{s0} has disappeared from the column of the basic indexes.

3. All the primal basic variables depend upon the value of the artificial constraint M.

4. The primal basic solution is not feasible even for the enlarged artificial problem.

This configuration calls for the application of the dual simplex algorithm, keeping in mind that M is a large positive number.

Initial Primal Tableau
(First Step of Artificial Constraint Algoritm)

T_1					Z	x_1	x_2	x_3	x_{s1}	x_{s2}	x_{s0}	sol	BI
1	0	−3	:	0	0	:	1	−1	3	1	0	0	: −1	x_{s1}
0	1	−1	:	0	0	:	−1	2	1	0	1	0	: 3	x_{s2}
0	0	1	:	0	0	:	1	1	[1]	0	0	1	: M	x_{s0} →
0	0	4	:	1	1	:	−3	−2	−4	0	0	0	: 0	Z

↑

Intermediate Primal Tableau
(Dual Simplex Algorithm)

T_2				Z	x_1	x_2	x_3	x_{s1}	x_{s2}	x_{s0}	sol	BI
−1/2	0	0	: 0	0	:	[−2]	−4	0	1	0	−3	: $-1 - 3M$	x_{s1} →
−1	1	0	: 0	0	:	−2	1	0	0	1	−1	: $3 - M$	x_{s2}
1/2	0	1	: 0	0	:	1	1	1	0	0	1	: M	x_3
1/2	0	0	: 1	1	:	1	2	0	0	0	4	: $4M$	Z

↑

Intermediate Primal Tableau
(Dual Simplex Algorithm)

$$
T_3 \quad
\begin{bmatrix}
1 & 0 & 2 & : & 0 \\
0 & 1 & 5 & : & 0 \\
0 & 0 & -1 & : & 0 \\
\cdots \\
0 & 0 & 0 & : & 1
\end{bmatrix}
$$

Z	x_1	x_2	x_3	x_{s1}	x_{s2}	x_{s0}	sol	BI
0 :	1	2	0	$-1/2$	0	$3/2$:	$\frac{1}{2}+\frac{3}{2}M$	x_1
0 :	0	5	0	-1	1	2 :	$4+2M$	x_{s2}
0 :	0	$\boxed{-1}$	1	$1/2$	0	$-1/2$:	$-\frac{1}{2}-\frac{1}{2}M$	$x_3 \rightarrow$
1 :	0	0	0	$1/2$	0	$5/2$:	$-\frac{1}{2}+\frac{5}{2}M$	Z

(arrow under x_2)

Intermediate Primal Tableau
(Dual Simplex Algorithm)

$$
T_4 \quad
\begin{bmatrix}
1 & 1 & 0 & : & 0 \\
0 & -2 & 0 & : & 0 \\
0 & 1 & 1 & : & 0 \\
\cdots \\
0 & 5 & 0 & : & 1
\end{bmatrix}
$$

Z	x_1	x_2	x_3	x_{s1}	x_{s2}	x_{s0}	sol	BI
0 :	1	0	2	$1/2$	0	$1/2$:	$-\frac{1}{2}+\frac{1}{2}M$	x_1
0 :	0	0	5	$3/2$	1	$\boxed{-1/2}$:	$\frac{3}{2}-\frac{1}{2}M$	$x_{s2} \rightarrow$
0 :	0	1	-1	$-1/2$	0	$1/2$:	$\frac{1}{2}+\frac{1}{2}M$	x_2
1 :	0	0	0	$1/2$	0	$5/2$:	$-\frac{1}{2}+\frac{5}{2}M$	Z

(arrow under x_{s0})

Final Tableau

Z	x_1	x_2	x_3	x_{s1}	x_{s2}	x_{s0}	P sol	BI
0 :	1	0	7	2	1	0 :	1	x_1
0 :	0	0	-10	-3	-2	1 :	$-3+M$	x_{s0}
0 :	0	1	4	1	1	0 :	2	x_2
1 :	0	0	25	8	5	0 :	7	Z
D sol	y_{s1}	y_{s2}	y_{s3}	y_1	y_2	y_0	R	

The final tableau is optimal because it exhibits primal and dual feasible solutions for the original problem. In fact, in this final tableau all the original variables have been freed from their dependence upon the value of the artificial constraint. The row corresponding to x_{s0} can be disregarded.

A check of the computations is always appropriate. First, let us identify the optimal solutions:

primal solution $= \{x_1 = 1,\ x_2 = 2,\ x_3 = 0,\ x_{s1} = 0,\ x_{s2} = 0\}$

dual solution $= \{y_{s1} = 0,\ y_{s2} = 0,\ y_{s3} = 25,\ y_1 = 8,\ y_2 = 5\}$

maximum Z $= 3(1) + 2(2) = 7 = -1(8) + 3(5) = $ minimum R

The optimal bases are:

$$\text{primal basis} = \begin{bmatrix} 1 & -1 \\ -1 & 2 \end{bmatrix} \qquad \text{dual basis} = \begin{bmatrix} 1 & -1 & 0 \\ -1 & 2 & 0 \\ 3 & 1 & -1 \end{bmatrix}$$

In general, the artificial constraint algorithm also leads to a two-phase approach. Phase I would be concerned with determining whether the original problem has a feasible solution. When the slack artificial variable x_{s0} reenters the primal basis, the original primal variables are liberated from the dependence upon the M value. This constitutes the end of phase I. Phase II is concerned with discovering whether it is possible to make the original primal basic solution feasible with a continuing use of the dual simplex algorithm. When a primal feasible solution is achieved, that solution is also optimal because the dual solution has been kept feasible from the beginning of the computations. In the above very simple numerical example it just happens that the original primal solution is feasible at the end of phase I.

To point out a further aspect of symmetric duality, it is interesting to note that the artificial constraint algorithm terminates when the vector associated with the artificial slack variable x_{s0} reenters the basis, while the artificial variable algorithm terminates when all the vectors associated with artificial variables x_{ai} exit the basis.

11.2   Termination of the Artificial Constraint Algorithm

As for the three other algorithms studied in chapters 7, 8, and 10, the artificial constraint algorithm can end up at three different statuses: at a finite optimal solution to the original problem, at an unbounded solution, and at an infeasible solution.

Chapter 12

The Diet Problem Revisited

12.1 The Simplex Solution of Stigler's Problem

In previous chapters we have acquired the ability to use the family of simplex algorithms for solving any linear programming problem. We also have developed the knowledge of how to interpret economically every component of the optimal solutions. It is time, therefore, to reconsider Stigler's diet problem, as promised in section 1.3, and to compute its true optimal solution.

Briefly, recall that Stigler formulated a relevant linear programming problem at a time when nobody knew how to solve it. Its mere formulation must be regarded as the impressive product of a trailblazing mind. For implementing his idea of a least-cost diet, Stigler had to choose a combination of foods from a list of 77 items (table 1.2 and table 12.3) to satisfy the minimal requirement level of 9 nutrients, as presented in table 1.1. He soon realized that many of the foods in the original list uniformly contained less nutritive value than others per dollar of expenditure. After he eliminated such foods, 15 items remained, as reported in table 1.2. Still, in 1945, the problem of selecting a least-cost diet was far from its solution. It is enlightening to read Stigler's own report (op. cit, p. 310):

> By making linear combinations of various commodities it is possible to construct a composite commodity which is superior in all respects to some survivor, and by this process the list of eligible commodities can be reduced at least to 9 [starred in table 1.2]. The nutritive values of each of these commodities is then expressed in terms of days' supply of requirements. Various

combinations of commodities were used to fulfill certain nutrient requirements, and the one finally chosen is presented in table 1.3. There is no reason to believe that the cheapest combination was found, for only a handful of the 510 possible combinations of commodities were examined. On the other hand the annual cost could not have been reduced by more than a few dollars by a better selection from these commodities.

As it turned out, George Stigler was overly cautious: he missed the annual minimum cost by a mere 27 cents! The optimal solution was computed using the information in tables 1.1 and 1.2 arranged to fit the specification of a diet problem such as

$$\text{minimize } TADC \quad = \quad \text{total annual diet cost}$$

$$\text{subject to} \qquad \text{nutrient supply} \geq \text{nutrient requirements}$$

or, in more explicit form,

$$\text{minimize } TADC = 365 \times (\$ \text{ food prices}) \times (\text{lb of food})$$

subject to

$$365 \times \frac{\text{nutrient supply}}{\text{lb of food}} \times (\text{lb of food}) \geq 365 \times \left(\begin{array}{c} \text{daily nutrient} \\ \text{requirements} \end{array} \right)$$

Using a general symbolic notation, the diet problem can be stated as

Primal minimize $TFC = \sum_{j=1}^{n} p_j x_j$

$$\text{subject to} \qquad \sum_{j=1}^{n} (ns)_{ij} x_j \geq (nr)_i, \qquad i = 1, \ldots, m$$

$$x_j \geq 0, \qquad j = 1, \ldots, n$$

where p_j is the price of the jth food, $(nr)_i$ is the ith nutrient requirement, $(ns)_{ij}$ is the supply of the ith nutrient contained in one unit of the jth food, and x_j is the level of the jth food.

Table 12.1. The Optimal Primal and Stigler's Solutions of the Diet Problem

Food	Units	Optimal Solution		Stigler's Solution	
		Daily	Annual	Daily	Annual
Wheat flour	lb	0.820	299.29	1.014	370.00
Evaporated milk	cans	...	...	0.156	57.00
Liver (beef)	lb	0.007	0.69	...	...
Cabbage	lb	0.303	110.63	0.304	111.00
Spinach	lb	0.062	22.56	0.063	23.00
Navy beans	lb	1.034	377.55	0.781	285.00
Total cost	$	0.1087	39.66	0.1094	39.93

The corresponding dual problem is

Dual maximize $VNR = \sum_{i=1}^{m} (nr)_i y_i$

subject to $\sum_{i=1}^{m} (ns)_{ji} y_i \leq p_j, \qquad j = 1, \ldots, n$

$$y_i \geq 0, \qquad i = 1, \ldots, m$$

The economic interpretation of the dual suggests that the diet problem can be seen as maximizing the value of the nutrient requirements (VNR) subject to the condition that the imputed marginal value of each food (the sum of all nutrient supplies times the corresponding dual variables) be less than or equal to its market price.

The optimal primal solution of the diet problem and Stigler's solution are reported in table 12.1. They are remarkably similar, and Stigler (and his legion of research assistants) must be congratulated. The calculation of the optimal solution on a medium-size computer required only a few seconds, while it took Stigler's assistants 120 man-days of work to compute their solution on desk calculators.

From table 12.1 we now know that in 1939 an adequate daily diet could have cost 11 cents, corresponding to 40 dollars per year, definitely a small sum even in those troubled times. On the other hand, wheat flour, navy beans, and cabbage in the required amounts would have thoroughly taxed the digestive system rather than the pocketbook of the most thrifty individual. Stigler was aware of the lack of palatability of his diet and never proposed it for implementation. In those times of food scarcity and debate over the meaning of good nutrition, however, he demonstrated the

Table 12.2. The Dual Solution of the Diet Problem (cents)

Daily Nutrient Requirements	Imputed Values	Nonbasic Foods	Opportunity Costs
Calories (000 calories)	0.877	Evaporated milk	0.293
Protein (grams)	0.000	Oleomargarine	12.527
Calcium (grams)	3.174	Cheese	5.685
Iron (milligrams)	0.000	Green beans	4.365
Vitamin A (000 I.U.)	0.040	Onions	1.997
Thiamine (milligrams)	0.000	Potatoes	0.761
Riboflavin (milligrams)	1.636	Sweet potatoes	1.785
Niacin (milligrams)	0.000	Dried peaches	11.904
Ascorbic acid (milligrams)	0.014	Dried prunes	6.001
		Lima beans	0.918

possibility of specifying and computing a least-cost adequate diet.

For our purpose, the diet problem and its solution present the opportunity for demonstrating the ability of linear programming to deal with real-life decision problems. The main virtue of the method consists in the ability to process and synthesize a large amount of information to guide the actions of the decision maker. One portion of this information that should never be overlooked is given by the dual solution. Stigler did not have any conceptual framework to obtain imputed values of the nutrient requirements, whereas we now know that the most appealing information for an economist comes from the dual problem. The dual solution of the diet problem is presented in table 12.2.

Five constraints are binding, as indicated by the corresponding positive dual variables: calories, calcium, vitamin A, riboflavin, and ascorbic acid. This is the reason why only five foods enter the primal solution at positive levels. The magnitude of the dual variables indicates that the calcium and riboflavin requirements are the most difficult to meet. The higher the imputed value of a constraint, the more costly it is to meet its requirements. This information derived from the dual solution is crucial for knowing which constraint to relax first in case the problem permits it. One thousand calories are valued at 0.877 cents. One gram of calcium is valued at 3.174 cents, while 1 milligram of riboflavin has an imputed cost of 1.636 cents. Vitamin A and ascorbic acid show that their imputed costs are close to zero. Proteins, iron, thiamine, and niacin definitely have a value of zero. This means simply that the combination of foods selected by the linear programming solution contains these nutrients in excess of the prescribed requirements, and one should not be overly concerned with the problem of meeting them as long as one can eat according to the LP solution. In other words, the protein, iron, thiamine, and niacin requirements do not represent effective constraints.

The opportunity costs of the foods that did not enter the basic optimal solution are a measure of unnecessary costliness with respect to the goal of an adequate diet. The price of evaporated milk, for example, is 6.7 cents per can (table 1.2). Its opportunity cost is 0.293 cents. This means that if the price could diminish to 6.407 cents per can, evaporated milk could enter the optimal solution while maintaining the same overall cost of the diet. Notice that in a minimization problem the opportunity cost must be interpreted as $(c_j - z_j)$ rather than $(z_j - c_j)$, as for a maximization objective. Hence, in the case of evaporated milk, the income foregone is $z_j = c_j - 0.293 = 6.7 - 0.293 = 6.403$.

12.2 The Optimal Diet from 77 Food Categories

The optimal diet presented in section 12.1 was computed using the information about the nutritive value contained in 15 different food categories as reported in table 1.2. Stigler, however, began with a collection of 77 foods. The lack of algorithms and computing facilities for handling medium- and large-size problems dictated the reduction of the initial list to the 15 and, then, 9 foods reported in table 1.2. Therefore, a legitimate curiosity lingers regarding the optimal solution computed directly from the information about all 77 foods. Could the diet's cost have been further reduced if all 77 foods had been considered? While the answer to this question in 1945 carried a prohibitive computational burden, it requires no essential cost today. The information about the nutritive value of all the foods available to Stigler (except for the 15 foods already reported in table 1.2) is given in table 12.3.

It is rather amazing to discover that the optimal primal and dual solutions using all 77 foods are identical to those reported in tables 12.1 and 12.2. This is a further indication of Stigler's ability to capture the essence of the diet problem.

12.3 A Multiperiod Least-Cost Diet Problem

Stigler's annual diet suffers from lack of palatability, which really means from lack of food variety over a year of time. Regarded simply as a daily diet, however, it was not entirely implausible for a lower-income adult living in the United States in the year 1939. This specification helps to focus attention on the fact that the notion of "palatability" is loaded with cultural

Table 12.3. Nutritive Values of Common Foods per Dollar of Expenditure, August 1939

Commodity	Unit	Price (cents)	Edible Weight per $ (grams)	Calories (1,000)	Protein (grams)	Calcium (grams)	Iron (mg)	Vit. A 1,000 (I.U.)	Thiamine (mg)	Riboflavin (mg)	Niacin (mg)	Ascorbic Acid (mg)
16. Macaroni	1 lb	14.1	3,217	11.6	418	0.7	54	—	3.2	1.9	68	—
17. Wheat cereals	28 oz	24.2	3,280	11.8	377	14.4	175	—	14.4	8.8	114	—
18. Corn flakes	8 oz	7.1	3,194	11.4	252	0.1	56	—	13.5	2.3	68	—
19. Corn meal	1 lb	4.6	9,861	36.0	897	1.7	99	30.9	17.4	7.9	106	—
20. Hominy grits	24 oz	8.5	8,005	28.6	680	0.8	80	—	10.6	1.6	110	—
21. Rice	1 lb	7.5	6,048	21.2	460	0.6	41	—	2.0	4.8	60	—
22. Rolled oats	1 lb	7.1	6,389	25.3	907	5.1	341	—	37.1	8.9	64	—
23. White bread	1 lb	7.9	5,742	15.0	488	2.5	115	—	13.8	8.5	126	—
24. Whole wheat bread	1 lb	9.1	4,985	12.2	484	2.7	125	—	13.9	6.4	160	—
25. Rye bread	1 lb	9.2	4,930	12.4	439	1.1	82	—	9.9	3.0	66	—
26. Pound cake	1 lb	24.8	1,829	8.0	130	0.4	31	18.9	2.8	3.0	17	—
27. Soda crackers	1 lb	15.1	3,004	12.5	288	0.5	50	—	—	—	—	—
28. Milk	1 qt	11.0	8,867	6.1	310	10.5	18	16.8	4.0	16.0	7	177
29. Butter	1 lb	30.8	1,473	10.8	9	0.2	3	44.2	—	0.2	2	—
30. Eggs	1 dozen	32.6	1,857	2.9	238	1.0	52	18.6	2.8	6.5	1	—
31. Cream	1/2 pt	14.1	1,689	3.5	49	1.7	3	16.9	0.6	2.5	—	17
32. Peanut butter	1 lb	17.9	2,534	15.7	661	1.0	48	—	9.6	8.1	471	—
33. Mayonnaise	1/2 pt	16.7	1,198	8.6	18	0.2	8	2.7	0.4	0.5	—	—
34. Crisco	1 lb	20.3	2,234	20.1	—	—	—	—	—	—	—	—
35. Lard	1 lb	9.8	4,628	41.7	—	—	—	0.2	—	0.5	5	—
36. Sirloin steak	1 lb	39.6	1,145	2.9	166	0.1	34	0.2	2.1	2.9	69	—

Table 12.3. (Continued)

Commodity	Unit	Price (cents)	Edible Weight per $ (grams)	Calories (1,000)	Protein (grams)	Calcium (grams)	Iron (mg)	Vit. A 1,000 (I.U.)	Thiamine (mg)	Riboflavin (mg)	Niacin (mg)	Ascorbic Acid (mg)
37. Round steak	1 lb	36.4	1,246	2.2	214	0.1	32	0.4	2.5	2.4	87	–
38. Rib roast	1 lb	29.2	1,553	3.4	213	0.1	33	–	–	2.0	–	–
39. Chuck roast	1 lb	22.6	2,007	3.6	309	0.2	46	0.4	1.0	4.0	120	–
40. Plate	1 lb	14.6	3,107	8.5	404	0.2	62	–	0.9	–	–	–
41. Leg of lamb	1 lb	27.6	1,643	3.1	245	0.1	20	–	2.8	3.9	86	–
42. Lamb chops (rib)	1 lb	36.6	1,239	3.3	140	0.1	15	–	1.7	2.7	54	–
43. Pork chops	1 lb	30.7	1,477	3.5	196	0.2	30	–	17.4	2.7	60	–
44. Pork loin roast	1 lb	24.2	1,874	4.4	249	0.3	37	–	18.2	3.6	79	–
45. Bacon	1 lb	25.6	1,772	6.7	152	0.2	23	–	1.8	1.8	71	–
46. Ham (smoked)	1 lb	27.4	1,655	10.4	212	0.2	31	–	9.9	3.3	50	–
47. Salt Pork	1 lb	16.0	2,835	18.8	164	0.1	26	–	1.4	1.8	–	–
48. Roasting chicken	1 lb	30.3	1,497	1.8	184	0.1	30	0.1	0.9	1.8	68	46
49. Veal cutlets	1 lb	42.3	1,072	1.7	156	0.1	24	–	1.4	2.4	57	–
50. Salmon, pink (can)	16 oz	13.0	3,489	5.8	705	6.8	45	3.5	1.0	4.9	209	–
51. Apples	1 lb	4.4	9,072	5.8	27	0.5	36	7.3	3.6	2.7	5	544
52. Bananas	1 lb	6.1	4,982	4.9	60	0.4	30	17.4	2.5	3.5	28	498
53. Lemons	1 doz	26.0	2,380	1.0	21	0.5	14	–	0.5	–	4	952
54. Oranges	1 doz	30.9	4,439	2.2	40	1.1	18	11.1	3.6	1.3	10	1,998
55. Carrots	1 bunch	4.7	6,080	2.7	73	2.8	43	188.5	6.1	4.3	89	608
56. Celery	1 stalk	7.3	3,915	0.9	51	3.0	23	0.9	1.4	1.4	9	313
57. Lettuce	1 head	8.2	2,247	0.4	27	1.1	22	112.4	1.8	3.4	11	449

Table 12.3. (Continued)

Commodity	Unit	Price (cents)	Edible Weight per $ (grams)	Calories (1,000)	Protein (grams)	Calcium (grams)	Iron (mg)	Vit. A 1,000 (I.U.)	Thiamine (mg)	Riboflavin (mg)	Niacin (mg)	Ascorbic Acid (mg)
58. Peaches (can)	no 2.5	16.8	4,894	3.7	20	0.4	10	21.5	0.5	1.0	31	196
59. Pears (can)	no 2.5	20.4	4,030	3.0	8	0.3	8	0.8	0.8	0.8	5	81
60. Pineapple (can)	no 2.5	21.3	3,993	2.4	16	0.4	8	2.0	2.8	0.8	7	399
61. Asparagus (can)	no 2	27.7	1,945	0.4	33	0.3	12	16.3	1.4	2.1	17	272
62. Green beans (can)	no 2	10.0	5,386	1.0	54	2.0	65	53.9	1.6	4.3	32	431
63. Pork & beans (can)	16 oz	7.1	6,389	7.5	364	4.0	134	3.5	8.3	7.7	56	–
64. Corn (can)	no 2	10.4	5,452	5.2	136	0.2	16	12.0	1.6	2.7	42	218
65. Peas (can)	no 2	13.8	4,109	2.3	136	0.6	45	34.9	4.9	2.5	37	370
66. Tomatoes (can)	no 2	8.6	6,263	1.3	63	0.7	38	53.2	3.4	2.5	36	1,253
67. Tomato soup (can)	10.5 oz	7.6	3,917	1.6	71	0.6	43	57.9	3.5	2.4	67	862
68. Raisins, dried	15 oz	9.4	4,524	13.5	104	2.5	136	4.5	6.3	1.4	24	136
69. Peas, dried	1 lb	7.9	5,742	20.0	1,367	4.2	345	2.9	28.7	18.4	162	–
70. Coffee	1 lb	22.4	2,025	–	–	–	–	–	4.0	5.1	50	–
71. Tea	1/2 lb	17.4	652	–	–	–	–	–	–	2.3	42	–
72. Cocoa	8 oz	8.6	2,637	8.7	237	3.0	72	–	2.0	11.0	40	–
73. Chocolate	8 oz	16.2	1,400	8.0	77	1.3	39	–	0.9	3.4	14	–
74. Sugar	10 lb	51.7	8,773	34.9	–	–	–	–	–	–	–	–
75. Corn sirup	24 oz	13.7	4,966	14.7	–	0.5	74	–	–	–	5	–
76. Molasses	18 oz	13.6	3,752	9.0	–	10.3	244	–	1.9	7.5	146	–
77. Strawberry preserve	1 lb	20.5	2,213	6.4	11	0.4	7	0.2	0.2	0.4	3	–

Source: Stigler, G. "The Cost of Subsistence," *Journal of Farm Economics*, 1945, pp. 306-7. The first 15 foods are reported in table 1.2

and ethnic connotations. Furthermore, many food prices are subject to seasonal fluctuations in response to the relative scarcity of those foods and their substitutes. These considerations suggest that, if we desire to improve the model specification for an annual least-cost diet, it is important to take into account the cultural and culinary habits of the target population as well as food availability.

To begin with, it seems useful to formulate a weekly consumption plan. This unit of time is neither too short nor too long for our purposes, and it can accommodate changes in prices and food availabilities to a satisfactory approximation. Furthermore, it is not unusual to vary the diet by the week. Secondly, we notice that Stigler's solution as well as the true optimal solution contain only cabbage and spinach among the vegetables but no fresh fruits and neither red nor white meat. Prices of fruits and vegetables are highly seasonal and Stigler might have compiled an average price for these commodities. In such a case, there could be a time window when these prices fall well below the annual average level, and the commodities could profitably enter the optimal diet. Prices of meat and other durable food are subject to sales discounts from time to time and, although it is rather difficult to know exactly when these sales occur, for the sake of exemplifying the issues involved we assume that they follow a known pattern. With food sales there comes the need of food storage for a period of time longer than a week. We assume that cold storage capacity is limited both because it is costly and because food might gradually lose some nutritive and palatability value. The storage option provides the link between the purchase and consumption decisions during the entire horizon of the problem.

The multiperiod diet problem specified below still constitutes a rather simplified example, but it gives an idea of the articulated structure needed for obtaining more plausible results. For the sake of keeping the formulation within a manageable size, suppose a LP consultant is charged with finding a minimum-cost diet for three consecutive periods (weeks). In general, food prices change weekly either because of announced sales or for seasonal reasons. The consultant assumes that it is possible to take advantage of low food prices in certain weeks through purchases of food amounts that may supply the necessary nutrient requirements up to the third week. Food can be stored up to two weeks at the cost of operating the storage facility (a family-size freezer and refrigerator). Although not all foods need cold storage, we will assume so to keep the notation to a minimum.

We deal with vectors of prices and of commodity quantities purchased, consumed, and stored. All these vectors are represented by the corresponding boldface symbols.

Let $\mathbf{p}_t$ be the price vector of foods available for purchase at time t; $\mathbf{x}_t$ is the quantity vector of commodities purchased at time t; $\mathbf{x}_t^c$ is the vector of commodities consumed at time t; $\mathbf{x}_t^s$ is the vector of commodities stored at time t; $\mathbf{x}_t^{s1}$, $\mathbf{x}_t^{s2}$ are the vectors of commodities taken out of storage for consumption after the first and second week, respectively. A_t is the matrix of nutrient supplies evaluated at time t; $\mathbf{r}_t$ is the vector of nutrient requirements at time t and s_t is the storage capacity available at time t. The symbol $\mathbf{u}$ is a vector whose components are all unit elements, often called the sum vector.

The primal specification of this simplified multiperiod diet problem can then be stated as follows:

Primal   minimize $TC = \displaystyle\sum_{\tau=0}^{T} \mathbf{p}'_{t+\tau}\mathbf{x}_{t+\tau}$

subject to

$$
\begin{array}{llll}
\mathbf{x}_t - \mathbf{x}_t^c - \mathbf{x}_t^s & = & 0 & \text{material balance} \\
A_t\mathbf{x}_t^c & \leq & \mathbf{r}_t & \text{requirements at time } t \\
\mathbf{u}'\mathbf{x}_t^s & \leq & s_t & \text{storage at time } t \\
\mathbf{x}_t^s - \mathbf{x}_t^{s1} - \mathbf{x}_t^{s2} & = & 0 & \text{storage allocation}
\end{array}
$$

all vectors are nonnegative, $\tau = 0, \ldots T$

The first constraint simply stipulates that the amount of food purchased, $\mathbf{x}_t$, must be equal to the amount of food consumed, $\mathbf{x}_t^c$, and stored, $\mathbf{x}_t^s$. The second constraint is the nutrient requirement. The third constraint is the storage capacity. Finally, the fourth constraint allows the division of the amount of food stored at time t into two installments to be released for consumption in the first and second week following time t.

The unfolding of the constraints continues to exhibit an almost recursive structure for all the required periods. The dimensions of the problem grow rather rapidly, but the computing capabilities available today do not pose any serious obstacle to the solution of a large-size linear programming model. Two of the four constraints characterizing each period are accounting constraints, and they are necessary for transferring the portion of food in storage either to the next storage or to the next consumption. The complete specification of the constraints for the three-period example is presented in table 12.4.

All the vectors entering this primal formulation are nonnegative. The interpretation of the constraints of this expanded diet problem has already been given above for time t. The same interpretation is applicable to time $t+1$ and $t+2$. Different weeks are linked to each other through the amount

Table 12.4. The Primal Specification of a Multiperiod Diet Problem

Primal minimize $TC = \mathbf{p}'_t\mathbf{x}_t + \mathbf{p}'_{t+1}\mathbf{x}_{t+1} + \mathbf{p}'_{t+2}\mathbf{x}_{t+2}$

subject to

1. $\mathbf{x}_t - \mathbf{x}^c_t - \mathbf{x}^s_t = 0$

2. $A_t\mathbf{x}^c_t \geq \mathbf{r}_t$

3. $-\mathbf{u}'\mathbf{x}^s_t \geq -s_t$

4. $\mathbf{x}^s_t - \mathbf{x}^{s1}_t - \mathbf{x}^{s2}_t = 0$

5. $\mathbf{x}_{t+1} - \mathbf{x}^c_{t+1} - \mathbf{x}^s_{t+1} = 0$

6. $A_t\mathbf{x}^{s1}_t + A_{t+1}\mathbf{x}^c_{t+1} \geq \mathbf{r}_{t+1}$

7. $-\mathbf{u}'\mathbf{x}^{s2}_t - \mathbf{u}'\mathbf{x}^s_{t+1} \geq -s_{t+1}$

8. $\mathbf{x}^s_{t+1} - \mathbf{x}^{s1}_{t+1} - \mathbf{x}^{s2}_{t+1} = 0$

9. $\mathbf{x}_{t+2} - \mathbf{x}^c_{t+2} - \mathbf{x}^s_{t+2} = 0$

10. $A_t\mathbf{x}^{s2}_t + A_{t+1}\mathbf{x}^{s1}_{t+1} + A_{t+2}\mathbf{x}^c_{t+2} \geq \mathbf{r}_{t+2}$

11. $-\mathbf{u}'\mathbf{x}^{s2}_{t+1} - \mathbf{u}'\mathbf{x}^s_{t+2} \geq -s_{t+2}$

Table 12.5. The Dual Specification of a Multiperiod Diet Problem

Dual

$$\text{maximize} \quad VNR = \sum_{\tau=0}^{2} \mathbf{r}'_{t+\tau}\mathbf{y}^r_{t+\tau} - \sum_{\tau=0}^{2} \mathbf{s}_{t+\tau}\mathbf{y}^s_{t+\tau}$$

subject to

1. $\mathbf{z}_t \le \mathbf{p}_t$

2. $A'_t\mathbf{y}^r_t \quad -\mathbf{z}_t \le 0$

3. $-\mathbf{z}_t \quad -\mathbf{u}y^s_t \quad +\mathbf{w}_t \le 0$

4. $-\mathbf{w}_t \quad +A'_t\mathbf{y}^r_{t+1} \le 0$

5. $-\mathbf{w}_t \quad -\mathbf{u}y^s_{t+1} \quad +A'_t\mathbf{y}^r_{t+2} \le 0$

6. $\mathbf{z}_{t+1} \le \mathbf{p}_{t+1}$

7. $A'_{t+1}\mathbf{y}^r_{t+1} \quad -\mathbf{z}_{t+1} \le 0$

8. $-\mathbf{u}y^s_{t+1} \quad -\mathbf{z}_{t+1} \quad +\mathbf{w}_{t+1} \le 0$

9. $A'\mathbf{y}^r_{t+2} \quad -\mathbf{w}_{t+1} \le 0$

10. $-\mathbf{w}_{t+1} \quad -\mathbf{u}y^s_{t+2} \le 0$

11. $\mathbf{z}_{t+2} \le \mathbf{p}_{t+2}$

12. $A'_{t+2}\mathbf{y}^r_{t+2} \quad -\mathbf{z}_{t+2} \le 0$

13. $-\mathbf{z}_{t+2} \quad -\mathbf{u}y^s_{t+2} \le 0$

of food stored for subsequent consumption. For example, the nutrient requirements of the third week can be met either by food purchased and consumed in that week $(t+2)$ or by food purchased previously (t and $t+1$) and stored.

The dual formulation of the multiperiod diet problem maximizes the value of the nutrient requirements for the entire horizon minus the imputed cost of operating the storage facility. Let $\mathbf{z}_t$ be the vector of dual variables associated with the material balance constraint at time t; $\mathbf{y}_t^r$ is the vector of dual variables associated with the nutrient requirement constraints; y_t^s is the dual variable associated with the storage constraint; $\mathbf{w}_t$ is the vector of dual variables associated with the storage allocation constraint. Then, the expanded dual problem is as shown in table 12.5.

The economic interpretation of the dual constraints follows the equilibrium pattern of

$$\boxed{\text{marginal value} \leq \text{marginal cost}}$$

Constraints 1 and 2 can be conveniently added together to produce

$$A_t'\mathbf{y}_t^r \leq \mathbf{p}_t$$

which is exactly the familiar equilibrium relation. Constraints 1, 3, and 4 can also be added together to give

$$A_t'\mathbf{y}_{t+1}^r \leq \mathbf{u}y_t^s + \mathbf{p}_t$$

which says that the unit value of food purchased at time t but consumed at time $t + 1$, $A_t'\mathbf{y}_{t+1}^r$ must be less than or equal to the price at time t plus the marginal cost of storing it for one week, $\mathbf{u}y_t^s$. The interpretation of constraint 5 is similar, although it requires a slightly more involved rearrangement

$$
\begin{aligned}
A_t'\mathbf{y}_{t+2}^r &\leq \mathbf{u}y_{t+1}^s + \mathbf{w}_t \\
&\leq \mathbf{u}y_{t+1}^s + \mathbf{u}y_t^s + \mathbf{p}_t \qquad \text{from constraints 3 and 1}
\end{aligned}
$$

The unit value of food purchased at time t but consumed at time $t+2$, $A_t'\mathbf{y}_{t+2}^r$ must be less than or equal to the purchase price plus the cost of storing it for two weeks $(\mathbf{u}y_{t+1}^s + \mathbf{u}y_t^s)$. Analogous interpretation can be developed for the remaining constraints, taking into account the time of purchase and consumption.

Since this diet model is a problem involving time, it is possible to derive an estimate of the internal rate of return (irr) implied by the use of the storage facility. The dual variable y_t^s is a measure of the freezer-refrigerator user cost. The relationship between the same dual variable in two different periods can be stated as

$$y^s_{t+1} = (1 + irr)y^s_t$$

and, similarly,

$$y^s_{t+2} = (1 + irr)y^s_{t+1} = (1 + irr)^2 y^s_t$$

With knowledge of the dual variables derived from the solution of the multiperiod linear program, the computation of the internal rate of return can be accomplished by solving the above equations. For example, from the first equation

$$irr = y^s_{t+1}/y^s_t - 1$$

The solution of the second equation involves the positive square root of the dual variables ratio. With suitable information about food availability and prices during the various weeks, the above multiperiod least-cost diet model could provide a consumption plan more plausible than an average annual diet.

Chapter 13

Parametric Programming: Input Demand Functions

13.1 Introduction

The previous twelve chapters were devoted to the task of learning the following:

1. How to set up linear programming problems.
2. The duality relations that bind primal and dual formulations.
3. The economic interpretation of LP problems.
4. How to solve linear programming problems.

Suppose we succeeded in obtaining an optimal solution of the given linear programming problem. Let us reflect for a moment on the significance of this event: a great deal of effort was expended in achieving this optimal solution. Indeed, we had to conceive the problem; gather the relevant information about the technological parameters, a_{ij}, the net revenue coefficients, c_j, the amount of limiting resources, b_i; and, finally, solve the linear programming problem. This last operation involves coding the data according to a fashionable LP computer program, checking the input data, obtaining a solution, and checking the solution for consistency and meaning. Now suppose that, while performing the verification of the results, we discover that one coefficient was typed incorrectly. Do we need to discard the optimal solution just obtained and begin the computations over again? The answer is no. We need only to perform a *sensitivity analysis*, that is, an analysis of the way the optimal solution is sensitive to changes in any original coefficients. Thus, one view of *parametric programming* is to determine the changes in the optimal basis and solution brought about by some variation in any of the problem's parameters. It concerns, therefore, the stability of the optimal basis, that is, the determination of whether it is necessary to change the optimal basis because some initial coefficient has changed. This viewpoint of parametric programming is useful but is not the most interesting.

The second viewpoint of parametric programming involves the role of economists in a more organic manner. Economists are ultimately interested in deriving demand and supply relations for the commodities of interest, inputs and outputs. With these relations they are able to articulate economic propositions involving equilibrium prices and quantities, surpluses, and shortages. Hence, the discussion of LP presented in previous chapters can be regarded as a preliminary work setting the stage for a more relevant analysis: the derivation of demand and supply functions for any commodity.

13.2   Derivation of Input Demand Functions

What are input derived demands, and how are they obtained? In the theory of the firm, the principal objective of the competitive entrepreneur is assumed to be that of maximizing profit (net revenue) subject to the prevailing technology, prices, and input availabilities. This is the economic framework adopted throughout this book. We present below a brief description of the logical process leading to the derivation of input demand functions. Let

$x \equiv f(b_1, \ldots, b_m)$ be a concave production function for a single output x using m inputs, $b_1, \ldots, b_m$

$p \equiv$ price of output x

$r_i \equiv$ price of the ith input

The profit objective of the competitive firm is then defined as the maximization of revenue minus costs, that is,

$$\text{maximize } \pi = px - (r_1 b_1 + r_2 b_2 + \ldots + r_m b_m)$$
$$= pf(b_1, \ldots, b_m) - \sum_{i=1}^{m} r_i b_i$$

In order to derive the optimal quantities of inputs that allow the maximization of the entrepreneur's profit (via the production of the optimal output quantity), we must differentiate the profit equation with respect to the inputs, set the corresponding derivatives equal to zero, and solve the resulting system of equations for the optimal quantities of inputs in terms of the parameters of the problem, that is, output and input prices. This set of equations is called the system of first-order conditions indicated as follows:

$$\frac{\partial \pi}{\partial b_1} = p\frac{\partial f(b_1,\ldots,b_m)}{\partial b_1} - r_1 \overset{\text{set}}{=} 0$$

$$\cdots \qquad \cdots \cdots \cdots \qquad \cdots \qquad \cdots$$

$$\frac{\partial \pi}{\partial b_m} = p\frac{\partial f(b_1,\ldots,b_m)}{\partial b_m} - r_m \overset{\text{set}}{=} 0$$

Notice that $\partial f/\partial b_i$ is the marginal product (MP) of the ith input. Hence, pMP_i is the value marginal product (VMP_i) of the ith input. Each of the above equations can, therefore, be stated as

$$VMP_i(b_1,\ldots,b_m) = r_i, \qquad i = 1,\ldots,m$$

Their meaning is that, for achieving maximum profit, the entrepreneur must choose the combination of inputs $b_1,\ldots,b_m$ that equates the value marginal product of each input to its corresponding price. The solution of the above system of equations, when it exists, expresses the input quantities $b_1,\ldots,b_m$ as a function of the prices $p,r_1,\ldots,r_m$, say $b_i = g_i(p,r_1,\ldots,r_m)$, $i = 1,\ldots,m$. These m functions are called the *derived demand functions* for inputs because the entrepreneur, in the quest for maximizing profits, wishes to demand those optimal input quantities on the market. The above equations, then, expressing prices as functions of input quantities are called *inverse input demand functions.* Figure 13.1 represents a plausible *inverse* demand function for the ith input. By changing the input quantity b_i (with every other input quantity maintained at the same level or, *ceteris paribus,* in economic terminology), it is possible to trace out the inverse input demand function of the ith input on a two-dimensional diagram. This is exactly what we wish to do in order to gain a better understanding of the characteristics of each input demand function. If we were to change two or more input quantities (or two or more prices), the result would be much more complex and difficult to describe. This level of complexity is the main reason for appealing to the *ceteris paribus* principle. Similarly, a *ceteris paribus* diagram of the ith direct input demand function $b_i = g_i(p,r_1,\ldots,r_m)$ is presented in figure 13.2.

No essential difference exists between the two relations illustrated in figures 13.1 and 13.2. In both cases, the input demand functions slope downward. The only difference is given by the axes of the two diagrams, which are inversely labeled. This is the reason for calling one of these two relation the inverse input demand function.

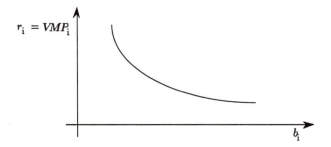

Figure 13.1. The inverse demand function of the ith input.

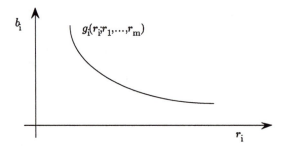

Figure 13.2. The direct demand function for the ith input.

13.3 Connection with Linear Programming

The discussion of the previous section has a direct connection with linear programming. When using a linear programming framework to represent a firm's problem, the coefficient c_j in the objective function can be interpreted as the output price (or marginal net revenue, MR) of the jth commodity; the constraint b_i is the limiting quantity of the ith input; the dual variable y_i represents the imputed (shadow) price of that limiting input, which, for an optimum solution, must be equal to its value marginal product, as discussed in section 13.2. This last relation is clearly identified by the process of computing dual variables in linear programming. From section 7.3, the ith dual variable can be stated as

$$y_i = VMP_i = MR \times MP_i$$

$$= \begin{bmatrix} \text{prices of} \\ \text{activities in the} \\ \text{production plan} \end{bmatrix} \times \begin{bmatrix} MP \text{ of the } i\text{th input with} \\ \text{respect to the activities} \\ \text{in the production plan} \end{bmatrix} = \mathbf{c}'_B \mathbf{mrtt}_{si}$$

The symbol y_i, therefore, corresponds to the input price r_i in the framework of the theory of the competitive firm presented in section 13.2. This discussion also makes it clear that in linear programming we deal with inverse input demand functions, since the shadow price of the ith input, y_i, is part of the dual solution, while the corresponding input quantity b_i is to be regarded as under the direct control of the entrepreneur and it is given for the LP problem. Furthermore, notice that a linear programming framework admits the analysis of the interrelationships between multiple outputs (joint and nonjoint) and multiple inputs. To verify all these propositions, consider the following example:

Primal maximize $= 3x_1 + 8x_2 + 5x_3$

$$\text{dual variables}$$

$$\text{subject to} \quad x_1 + 4x_2 + x_3 \le 12 \equiv b_1 \quad y_1$$

$$6x_1 + 2x_2 + 3x_3 \le 18 \equiv b_2 \quad y_2$$

$$x_j \ge 0, \quad j = 1, 2, 3$$

The initial and the optimal tableaux are as follows:

Initial Primal Tableau

Z	x_1	x_2	x_3	x_{s1}	x_{s2}	sol	BI
0	1	4	1	1	0	12	x_{s1}
0	6	2	3	0	1	18	x_{s2}
1	-3	-8	-5	0	0	0	Z

First Optimal Tableau

Z	x_1	x_2	x_3	x_{s1}	x_{s2}	sol P	BI
0	$-3/10$	1	0	$3/10$	$-1/10$	$9/5$	x_2
0	$11/5$	0	1	$-1/5$	$2/5$	$24/5$	x_3
1	$28/5$	0	0	$7/5$	$6/5$	$192/5$	Z
D sol	y_{s1}	y_{s2}	y_{s3}	y_1	y_2	R	

From the optimal tableau:

primal solution $= \{x_1 = 0, x_2 = 9/5, x_3 = 24/5, x_{s1} = 0, x_{s2} = 0\}$

dual solution $= \{y_{s1} = 28/5, y_{s2} = 0, y_{s3} = 0, y_1 = 7/5, y_2 = 6/5\}$

primal basis $= [\mathbf{a}_2 \quad \mathbf{a}_3] = \begin{bmatrix} 4 & 1 \\ 2 & 3 \end{bmatrix}$

dual basis $= [-\mathbf{e}_{D1} \quad \mathbf{a}_{D1} \quad \mathbf{a}_{D2}] = \begin{bmatrix} -1 & 1 & 6 \\ 0 & 4 & 2 \\ 0 & 1 & 3 \end{bmatrix}$

Shadow prices as value marginal products. Let us verify that the dual variables y_1 and y_2 can be interpreted as value marginal product functions, as stated above: given that the optimal primal basis is composed of activities #2 and #3, the vector of prices associated with activities in the production plan is $\mathbf{c}'_B = [8 \quad 5]$; hence,

$$y_1 = MR \times MP_1 = \mathbf{c}'_B \mathbf{mrtt}_{s1} = [8 \quad 5] \begin{bmatrix} 3/10 \\ -1/5 \end{bmatrix} = \frac{7}{5}$$

$$y_2 = MR \times MP_2 = \mathbf{c}'_B \mathbf{mrtt}_{s2} = [8 \quad 5] \begin{bmatrix} -1/10 \\ 2/5 \end{bmatrix} = \frac{6}{5}$$

To show that the coefficients $(3/10)$ and $(-1/5)$ in the column corresponding to slack variable x_{s1} of the optimal tableau should be interpreted as marginal products of input #1, let us consider the first row of coefficients in the optimal tableau. This row can be read as an equation by simply reattaching the variable names to the respective coefficients, remembering that $x_1 = x_{s2} = 0$ (*ceteris paribus*):

$$x_2 + \frac{3}{10} x_{s1} = \frac{9}{5}$$

Recall also that the computation of a marginal product (section 7.2.2) requires taking the total differential of the production function and then the derivative with respect to the variable of interest. Therefore,

$$dx_2 + \frac{3}{10} dx_{s1} = 0$$

and

$$-\frac{dx_2}{dx_{s1}} = \frac{3}{10} \equiv MP_{x_2, b_1}$$

Since x_2 is interpreted as an output and x_{s1} as an input, we recognize that the derivative $-dx_2/dx_{s1}$ is the marginal product of input #1 relative to output #2. Similarly, from the second row of the optimal tableau, the marginal product of input #1 relative to output #3 can be computed as

$$x_3 - \frac{1}{5}x_{s1} = \frac{24}{5}$$

$$dx_3 - \frac{1}{5}dx_{s1} = 0$$

and

$$-\frac{dx_3}{dx_{s1}} = -\frac{1}{5} \equiv MP_{x_3, b_1}$$

Analogous computations can be shown for input #2. The column of coefficients under x_{s2} in the optimal tableau indicates the marginal products of input #2 relative to outputs #2 and #3:

$$MP_{x_2, b_2} \equiv -\frac{dx_2}{dx_{s2}} = -\frac{1}{10}$$

$$MP_{x_3, b_2} \equiv -\frac{dx_3}{dx_{s2}} = \frac{2}{5}$$

Marginal products can be negative, as the above example illustrates. This is another interesting feature of linear programming admitting the production of an optimal output mixture with negative marginal products for individual outputs, as long as all the negative marginal products do not refer to the same output.

13.4 The Derived Demand for Input #1

We are ready to discuss the parametric computation of the derived demand function for, say, input #1. Because of the shape, as illustrated in this section, the function is often also called a *schedule*. The two names are equivalent.

With the information contained in the optimal tableau, it is possible to begin drawing two important diagrams relating the optimal revenue (figure 13.3) and the shadow price y_1 (figure 13.4) to the quantity of input #1. These figures represent very incomplete revenue and inverse input demand diagrams, illustrating that by merely solving a linear programming problem, all we can do is plot a single point of either function.

The optimal tableau also indicates that the optimal basis is formed by activities #2 and #3. It is possible, therefore, to draw a third important diagram as in figure 13.5, where the vectors a_2 and a_3, representing the optimal basis, are connected with an arc.

The relevant question now is: What happens to the optimal basis and to the primal and dual solutions if the initial availability of input #1 is increased?

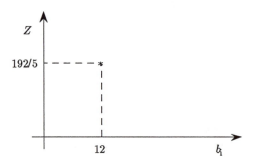

Figure 13.3. The revenue function after the first optimal tableau.

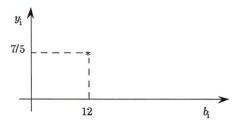

Figure 13.4. The inverse demand function for input #1.

13.4.1 Step 1: Increase in b_1 (the Rybczynski Theorem)

As b_1 increases (from $b_1 = 12$, with b_2 remaining at the initial level of 18), the vector of limiting resources **b** in the input space diagram rotates clockwise. A sufficient increase of input #1 makes the **b** vector coincident with the a_2 vector. Figure 13.5 also indicates that, for all increases in b_1 corresponding to a rotation of the **b** vector within the cone generated by a_2 and a_3, the optimal basis needs not to (and does not) change. As a direct consequence, the optimal dual solution also does not change. Only the optimal primal solution changes in relation to the increase in the availability of input #1. When the changing **b** vector becomes coincident with the activity vector a_2, the primal solution becomes degenerate (section 8.5). Further increases in the quantity of input #1 will make the current optimal basis $[a_2 \quad a_3]$ infeasible (the **b** vector will be outside the cone generated by a_2 and a_3). During all these variations, the opportunity costs $(z_j - c_j)$ have remained nonnegative, as they were in the optimal tableau. Thus, before attempting further increases in b_1, we need to apply the dual simplex algorithm, since we wish to maintain the feasibility of the primal solution.

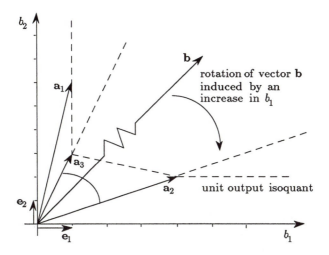

Figure 13.5. The increase of b_1 in the input requirement space.

The immediate conclusion of the preceding discussion is that parametric programming of the right-hand-side vector, as the increase in any limiting input is also called, begins with the determination by how much it is possible to increase input #1 without the need to change the optimal basis $[\mathbf{a}_2 \quad \mathbf{a}_3]$. In equivalent words, we wish to find out how the primal solution can change in relation to an increase of input #1, without the variables x_2 and x_3 losing their basic and feasible status. In symbolic notation, we wish to find a nonnegative increment Δb_1 of b_1 such that

$$\text{new } b_1 \quad = \quad \text{old } b_1 + \Delta b_1$$

and

$$\text{new } x_2 \quad \geq \quad 0$$
$$\text{new } x_3 \quad \geq \quad 0$$

This task can be accomplished by focusing on the optimal tableau and rewriting the first two rows of coefficients in the form of the following system of equations, keeping $x_1 = x_{s2} = 0$:

(N) $\quad$ new $x_2 = x_2 + (3/10)x_{s1} = 9/5 + (3/10)(\Delta b_1 \equiv x_{s1}) \geq 0$
$\qquad$ new $x_3 = x_3 - (1/5)x_{s1} = 24/5 - (1/5)(\Delta b_1 \equiv x_{s1}) \geq 0$

since, by definition, a slack variable is the difference between input availability and input utilization. Hence, a convenient notation for the slack variable x_{s1} is also

$$x_{s1} = b_1 - (x_1 + 4x_2 + x_3) \equiv \Delta b_1$$

This more suggestive notation will help to emphasize the direction and the meaning of the analysis. The feasibility of the new levels of x_2 and x_3 can now be maintained through the solution of the (N) system of inequalities by means of the familiar approach represented by the minimum ratio criterion:

$$\text{new } x_2 = 9/5 + (3/10)\Delta b_1 \geq 0$$
$$\text{new } x_3 = 24/5 - (1/5)\Delta b_1 \geq 0$$

We have encountered similar inequalities in the development of the feasibility criterion for the primal simplex algorithm (section 7.2.1). Feasibility of the new levels of x_2 and x_3, therefore, requires that Δb_1 be chosen according to the minimum ratio criterion where the denominator of the ratio must always be a positive number. According to this criterion, there is only one admissible ratio in the above numerical example and it is given by the second inequality, where

$$24/5 \geq (1/5)\Delta b_1$$

and, therefore,

$$\Delta b_1 = \frac{24/5}{1/5} = 24$$

According to the other inequality relation, $9/5 + (3/10)\Delta b_1 \geq 0$, Δb_1 could equal $+\infty$ without causing the infeasibility of the new x_2, but determining the infeasibility of the new x_3. We conclude that b_1 can increase from the initial 12 units to 36 units without the need to change the optimal primal basis $[a_2 \quad a_3]$, since

$$\text{new } b_1 = \text{old } b_1 + \Delta b_1 = 12 + 24 = 36$$

The new optimal primal solution, corresponding to the new value of the input vector $\mathbf{b}$ (new $b_1 = 36$, new $b_2 = $ old $b_2 = 18$), is easily computed by substituting $\Delta b_1 = 24$ in the (N) system of inequalities:

$$\text{(N)} \qquad\qquad \text{new } x_2 = 9/5 + (3/10)24 = 9$$
$$\text{new } x_3 = 24/5 - (1/5)24 = 0$$

As anticipated, the new optimal primal solution is degenerate. This means that activity a_3 must leave the basis if we intend to continue our analysis for a further increase of input #1. The countercheck is provided

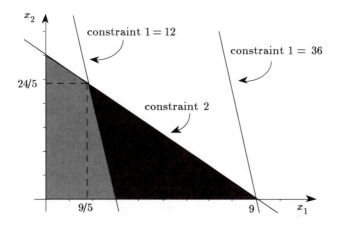

Figure 13.6. The Rybczynski theorem illustrated.

by selecting $\Delta b_1 = 25$. With this increment, the new primal solution would be: new $x_2 = 93/10$, but new $x_3 = -1/5$, an infeasible level.

The Rybczynski theorem. We observe that an increase in input #1 has had two opposite effects on the equilibrium output of commodities #2 and #3 as x_2 has increased and x_3 has decreased. Why did this happen? And what are the economic and technological implications of this event? To answer these questions, it is useful to look at the system of technological constraints in its original units (leaving out activity #1, since $x_1 = 0$):

$$4x_2 + x_3 = 12 + 24$$
$$2x_2 + 3x_3 = 18$$

We notice that activity a_2 uses input #1 more intensively than activity a_3, as $a_{21} = 4 > 1 = a_{31}$. The calculations carried out above under the profit maximization objective have shown that, with an increase of input #1, the level of x_2 has increased while that of x_3 has decreased. This behavior of the output mix is a consequence of a general proposition known in economics as the Rybczynski theorem, which says that, given that outputs use the same inputs, when one input is increased, the level of output associated with the activity that uses that resource more intensively will increase while the other activity level will decrease.

Figure 13.6 illustrates this proposition graphically by presenting the output space for (x_2, x_3). The original feasible region corresponding to $b_1 = 12$ and $b_2 = 18$ is lightly shaded. At the original equilibrium point

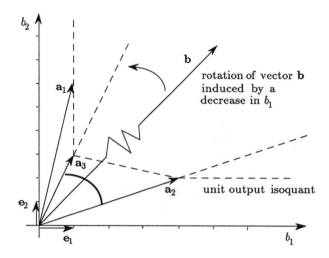

Figure 13.7. The decrease of b_1 in the input requirement space.

$x_2 = 9/5$ and $x_3 = 24/5$. The increase of b_1 from 12 to 36 shifts constraint #1 to the right, enlarging the feasible region and causing a new equilibrium at $x_2 = 9$ and $x_3 = 0$. Any increase of b_1 in the interval (12, 36) would have caused an increase in x_2 and a decrease in x_3, as the Rybczynski theorem states. Conversely, a decrease of input #1 will have the opposite effects on the level of output #2 and #3, as the next section will clarify.

13.4.2 Step 1: Decrease in b_1

Before discussing a further increase of input #1, it is convenient to determine the consequences of a *decrease* in the available initial quantity of the same input, $b_1 = 12$. The analytical machinery for this case is all in place. Consider, in fact, figure 13.7, which is a replica of figure 13.5 with the reversed rotation of the **b** vector to account for the decrease of b_1. As b_1 is decreased from the initial level of 12, the **b** vector rotates counterclockwise and there comes a point when it will coincide with the activity vector $\mathbf{a}_3$. At that point, the primal solution becomes degenerate and, if b_1 is further decreased, the same solution will become infeasible (**b** will be outside the cone generated by $\mathbf{a}_2$ and $\mathbf{a}_3$). The decrease of b_1 is symbolized as $-\Delta b_1$.

The same (N) system of inequalities as stated above will, then, be solved for the maximum allowed decrement of b_1:

$$(\text{N}) \qquad\qquad \text{new } x_2 = 9/5 + (3/10)(-\Delta b_1) \geq 0$$
$$\text{new } x_3 = 24/5 - (1/5)(-\Delta b_1) \geq 0$$

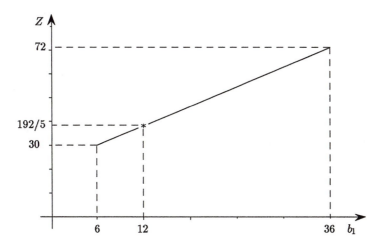

Figure 13.8. The revenue function after step 1.

The first inequality provides the solution of the problem with a value of $-\Delta b_1 = -6$, since the second inequality does not block $-\Delta b_1$ from assuming an infinite value. The new level of input #1 is now

$$\text{new } b_1 = 12 - 6 = 6$$

while b_2 remains at the same initial level of 18. The new primal solution corresponding to this decrement of input #1 is easily determined by replacing the value of $-\Delta b_1$ in the (N) system:

(N) $\qquad\qquad$ new $x_2 = 9/5 + (3/10)(-6) = 0$

$\qquad\qquad\qquad\qquad$ new $x_3 = 24/5 - (1/5)(-6) = 6$

A summary of the results achieved so far consists of the following findings:

1. Input #1 can vary between the levels of 6 and 36 units without the necessity of changing the current optimal basis $[\mathbf{a}_2 \quad \mathbf{a}_3]$.

2. Within the above interval of b_1, (6, 36), the dual variables remain constant at the initial optimal levels $\{y_1 = 7/5, \ y_2 = 6/5\}$.

3. With $b_1 = 36$, the optimal primal solution is $\{x_1 = 0, \ x_2 = 9, \ x_3 = 0, \ x_{s1} = 0, x_{s2} = 0\}$ and the corresponding value of the primal objective function is $Z = 72$.

4. With $b_1 = 6$, the optimal primal solution is $\{x_1 = 0, \ x_2 = 0, \ x_3 = 6, \ x_{s1} = 0, \ x_{s2} = 0\}$ and the corresponding value of the primal objective

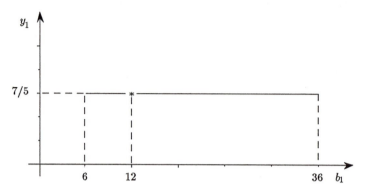

Figure 13.9. The derived demand for input #1 after step 1.

function is $Z = 30$.

This information allows us to expand the diagrams of the revenue and input demand functions as in figures 13.8 and 13.9.

13.4.3   Step 2: A Further Increase in b_1

The analysis of an increase in the initial level of b_1 continues with the routine question of what happens if input #1 is increased beyond the level of 36 units. We already know part of the answer. The primal solution will become infeasible (the **b** vector will lie outside the cone generated by a_2 and a_3). To avoid that event, we must change the optimal primal basis. This task requires using the dual simplex algorithm because, at $b_1 = 36$, the optimal primal solution is degenerate $\{x_2 = 9,\ x_3 = 0\}$ while the primal optimality criterion is satisfied (all opportunity costs are nonnegative). We continue, therefore, the simplex computations using the first optimal tableau with the new solution under $b_1 = 36$ replacing the original optimal solution.

First Optimal Tableau

$$
\begin{bmatrix} 1 & 3/2 & \vdots & 0 \\ 0 & -5 & \vdots & 0 \\ \cdots & \cdots & & \cdots \\ 0 & 7 & \vdots & 1 \end{bmatrix}
$$

	Z	x_1	x_2	x_3	x_{s1}	x_{s2}	sol	BI
	0 $\vdots$	$-3/10$	1	0	$3/10$	$-1/10$ $\vdots$	9	x_2
	0 $\vdots$	$11/5$	0	1	$\boxed{-1/5}$	$2/5$ $\vdots$	0	$x_3 \rightarrow$
	1 $\vdots$	$28/5$	0	0	$7/5$	$6/5$ $\vdots$	72	Z

$\uparrow$

Second Optimal Tableau (for $b_1 \uparrow$)

	Z	x_1	x_2	x_3	x_{s1}	x_{s2}	sol	BI
	0	3	1	$3/2$	0	$1/2$	9	x_2
	0	-11	0	-5	1	-2	0	x_{s1}
	1	21	0	7	0	4	72	Z
D sol	y_{s1}	y_{s2}	y_{s3}	y_1	y_2		R	

The slack activity e_1 of input #1 has replaced activity a_3 in the basis. The shadow price of input #1 is now zero. This suggests that, with the slack x_{s1} in the optimal basic solution, further increases of input #1 will augment only the unused quantity of input #1 without contributing to the production and revenue generating process. In equivalent words, when a surplus of an input exists, the corresponding value marginal product is equal to zero. By simple inspection of the second optimal tableau, we conclude that any increase of b_1 beyond the level of 36 units will leave everything (except x_{s1}) unchanged.

13.4.4 Step 2: A Further Decrease in b_1

The consequences of decreasing the quantity of input #1 below the level of 6 units can be determined through analogous reasoning. We know already that with $b_1 = 6$ and $b_2 = 18$, the optimal primal solution is $\{x_1 = 0, x_2 = 0, x_3 = 6, x_{s1} = 0, x_{s2} = 0\}$ and $Z = 30$. The basic components of the optimal solution are $(x_2 = 0, x_3 = 6)$ (consult section 13.4.2). Using again the first optimal tableau and replacing the old basic solution with the new one corresponding to $b_1 = 6$, we obtain the required change of basis through another application of the dual simplex algorithm:

First Optimal Tableau

$$\begin{bmatrix} -10 & 0 & \vdots & 0 \\ 4 & 1 & \vdots & 0 \\ 12 & 0 & \vdots & 1 \end{bmatrix}$$

	Z	x_1	x_2	x_3	x_{s1}	x_{s2}	sol	BI
	0	$-3/10$	1	0	$3/10$	$\boxed{-1/10}$	0	$x_2 \rightarrow$
	0	$11/5$	0	1	$-1/5$	$2/5$	6	x_3
	1	$28/5$	0	0	$7/5$	$6/5$	30	Z

$\uparrow$

Second Optimal Tableau (for $b_1 \downarrow$)

$$
\begin{array}{c|ccccccc|c}
 & Z & x_1 & x_2 & x_3 & x_{s1} & x_{s2} & \text{sol} & \text{BI} \\
\hline
 & 0 & 3 & -10 & 0 & -3 & 1 & 0 & x_{s2} \\
 & 0 & 1 & 4 & 1 & 1 & 0 & 6 & x_3 \\
\hline
 & 1 & 2 & 12 & 0 & 5 & 0 & 30 & Z \\
\end{array}
$$

D sol $\quad y_{s1} \quad y_{s2} \quad y_{s3} \quad y_1 \quad y_2 \quad R$

The shadow price of input #1 has jumped from the original $y_1 = 7/5$ to $y_1 = 5$ in correspondence to a decrease of the level of b_1 from 12 to 6. The current optimal basis is $[e_2 \quad a_3]$.

What happens, then, if we further reduce input #1 below 6 units? To answer the question, it is necessary to repeat the procedure developed in step 1. From the second optimal tableau associated with a decrease in b_1, we can write ($x_1 = x_2 = x_{s1} = 0$):

(N′) $\begin{aligned} \text{new } x_{s2} &= x_{s2} - 3(-\Delta b_1) = 0 - 3(-\Delta b_1) \geq 0 \\ \text{new } x_3 &= x_3 + (-\Delta b_1) = 6 + (-\Delta b_1) \geq 0 \end{aligned}$

From the second relation (in the first relation $\Delta b_1 = -\infty$)

$$-\Delta b_1 = -6$$

Thus, input #1 can be further reduced by 6 units to the level of zero units as:

$$\text{new } b_1 = \text{old } b_1 + (-\Delta b_1) = 6 - 6 = 0$$

and the current optimal basis $[e_2 \quad a_3]$ need not change. As a consequence, the dual variables do not change. The new optimal primal solution is derived from (N′):

(N′) $\begin{aligned} \text{new } x_{s2} &= 0 - 3(-6) = 18 \\ \text{new } x_3 &= 6 + (-6) = 0 \end{aligned}$

With no productive activities performed at positive levels, the revenue is also zero. The parametric analysis of input #1 is terminated, since negative quantities of an input are meaningless.

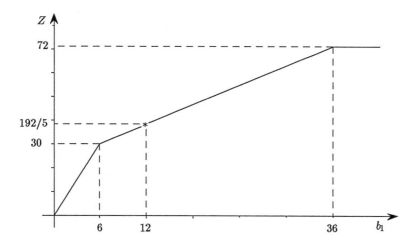

Figure 13.10. The revenue function after step 2.

It is now possible to complete the drawing of the revenue and inverse input demand diagrams as in figures 13.10 and 13.11. The values of the dual variable y_1 obtained in correspondence with either an increase or a decrease of input #1 allow the plotting of the revenue and input demand functions in their entirety. The derivatives, taken within the linear segments (with constant slope) of the revenue functions with respect to the input quantity b_1, result in the input demand schedule diagram of figure 13.11. As b_1 increases from 0, the revenue function becomes less steep. This decreasing slope is indicated by the dropping of the $VMP_1 = y_1$ schedule at exactly the same points of b_1 corresponding to the changes in slope of the revenue function. Because of the curvature exhibited in figure 13.10, the revenue function is said to be a concave function of the input quantities.

We summarize the results by stating that, within the intervals of b_1, corresponding to the same optimal basis, $\partial Z / \partial b_1 = y_1$, the imputed price of input #1. Thus, in linear programming the relationship between y_1 and b_1 is a nonincreasing function corresponding to the general economic idea of a demand function.

A final comment: the first step in a parametric analysis is often called sensitivity analysis. It determines how sensitive the optimal basis is to variations of input (constraints) quantities. Another term for the same step is equilibrium analysis. It should be clear that a far greater amount of information results from parametric programming compared to sensitivity analysis. The numerical computations of parametric programming are only marginally more demanding compared to sensitivity analysis.

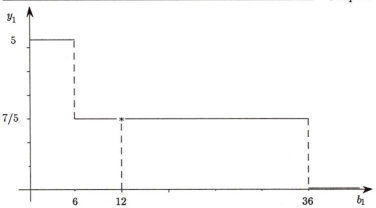

Figure 13.11. The derived demand for input #1 after step 2.

13.5   The Derived Demand for Input #2

To reinforce the rationale and understanding of the procedure for deriving an input demand function developed in the previous section, we will repeat here the entire process with respect to input #2. In so doing, we will have the opportunity to emphasize the notion that parametric programming, while requiring a low level of computations (mostly copying), is rich in information about the behavior of input demand relations.

Let us review the rationale for right-hand-side parametric programming. With such a name we refer to the parametric variation of one (or more) limiting input quantity, b_i, which appears on the right-hand side of the equality-inequality constraint. It should be clear that the same procedure can be used for any type of LP constraint, regardless of whether it represents an input, an output, or any other restriction.

We now intend to derive the (inverse) demand function for input #2. This means that we wish to modify the right-hand-side vector of limiting resources, **b**, obtaining a new vector that is equal to the old **b** except for one component, b_2. That is, we wish to find a new **b**, such that

$$\text{new } \mathbf{b} = \text{old } \mathbf{b} + \Delta\mathbf{b}$$
$$= \begin{bmatrix} 12 \\ 18 \end{bmatrix} + \begin{bmatrix} 0 \\ \Delta b_2 \end{bmatrix}$$

and the optimal primal basis need not to be changed.

13.5.1   Step 1: Increase and Decrease in b_2

A change in the vector of resources **b** will obviously modify the optimal quantities of outputs represented by the basic component of the solution $\mathbf{x}_B$. Hence, recalling that $\Delta b_2 = x_{s2}$, the relevant set of inequalities is

$$(\text{N}) \qquad \text{new } \mathbf{x}_B \;=\; \text{old } \mathbf{x}_B \quad \pm \; MRTT_s \times \Delta \mathbf{b} \;\geq\; 0$$

$$= \begin{bmatrix} 9/5 \\ 24/5 \end{bmatrix} \pm \begin{bmatrix} -1/10 \\ 2/5 \end{bmatrix} \Delta b_2 \;\geq\; 0$$

since we wish to change only b_2. The $\pm$ sign indicates that we wish to study the influence of both an increase (+) and a decrease (−) of b_2 on the optimal solution. To preserve the feasibility of the new solution (new $\mathbf{x}_B \geq 0$), Δb_2 must be chosen in such a way that its value will not cause any violation of the inequalities.

Increase in b_2. For an increase in b_2, the + sign is in operation, and from the first relation of (N) (the second inequality does not block Δb_2)

$$9/5 - (1/10)\Delta b_2 \geq 0$$

we compute $\Delta b_2 = 18$. The new configuration of the LP problem is now

$$\text{new } \mathbf{b} \;=\; \begin{bmatrix} 12 \\ 18 \end{bmatrix} \;+\; \begin{bmatrix} 0 \\ 18 \end{bmatrix} \;=\; \begin{bmatrix} 12 \\ 36 \end{bmatrix}$$

$$\text{new } \mathbf{x}_B \;=\; \begin{bmatrix} 9/5 \\ 24/5 \end{bmatrix} \;+\; \begin{bmatrix} -1/10 \\ 2/5 \end{bmatrix} 18 \;=\; \begin{bmatrix} x_2 = 0 \\ x_3 = 12 \end{bmatrix}$$

$$\text{new } Z \;=\; \begin{bmatrix} 8 & 5 \end{bmatrix} \begin{bmatrix} 0 \\ 12 \end{bmatrix} \;=\; 60$$

The new optimal primal solution is degenerate in x_2 (as expected, $x_2 = 0$) and a further increase of b_2 beyond 36 units requires a change of basis. This is accomplished by reproducing the first optimal tableau, replacing the old solution with the new solution and using the dual simplex algorithm for changing the basis.

First Optimal Tableau

$$
\begin{array}{c|ccccc|cc}
Z & x_1 & x_2 & x_3 & x_{s1} & x_{s2} & \text{sol} & \text{BI}\\
\hline
0 & -3/10 & 1 & 0 & 3/10 & \boxed{-1/10} & 0 & x_2 \rightarrow\\
0 & 11/5 & 0 & 1 & -1/5 & 2/5 & 12 & x_3\\
\hline
1 & 28/5 & 0 & 0 & 7/5 & 6/5 & 60 & Z
\end{array}
$$

The application of the dual simplex algorithm indicates that the surplus variable x_{s2} will enter the basis in the new tableau and, therefore, the corresponding dual variable y_2 will be zero. This means that any increase of input b_2 beyond the level of 36 units will be left unused by the current technology. The parametric analysis for an increase of b_2 terminates here with the realization that a little reasoning about the process can avoid many numerical computations.

Decrease in b_2. For a decrease in b_2, $-\Delta b_2 < 0$ (starting from the first optimal solution), the feasibility conditions stated in system (N) above indicate that the second relation is blocking the decrease of $-\Delta b_2$, that is,

$$24/5 + (2/5)(-\Delta b_2) \geq 0$$

and, therefore,

$$-\Delta b_2 = (-24/5)/(2/5) = -12$$

For a decrement of $\Delta b_2 = 12$ units, the new resource vector is

$$\text{new } \mathbf{b} = \begin{bmatrix} 12 \\ 18 \end{bmatrix} + \begin{bmatrix} 0 \\ -12 \end{bmatrix} = \begin{bmatrix} 12 \\ 6 \end{bmatrix}$$

the corresponding new primal solution is

$$\text{new } \mathbf{x}_B = \begin{bmatrix} 9/5 \\ 24/5 \end{bmatrix} + \begin{bmatrix} -1/10 \\ 2/5 \end{bmatrix}(-12) = \begin{bmatrix} x_2 = 3 \\ x_3 = 0 \end{bmatrix}$$

and the value of the objective function is

$$\text{new } Z = \begin{bmatrix} 8 & 5 \end{bmatrix}\begin{bmatrix} 3 \\ 0 \end{bmatrix} = 24$$

Again, the solution is degenerate, by construction, and a further decrease of b_2 below the level of 6 units requires a change of basis using the dual simplex algorithm. As for the increase, the new solution just computed replaces the old solution of the first optimal tableau.

First Optimal Tableau

$$
\begin{bmatrix} 1 & 3/2 & \vdots & 0 \\ 0 & -5 & \vdots & 0 \\ \cdots & \cdots & & \cdots \\ 0 & 7 & \vdots & 1 \end{bmatrix}
$$

	Z	x_1	x_2	x_3	x_{s1}	x_{s2}	sol	BI
	0 :	−3/10	1	0	3/10	−1/10 :	3	x_2
	0 :	11/5	0	1	$\boxed{-1/5}$	2/5 :	0	x_3 →
	1 :	28/5	0	0	7/5	6/5 :	24	Z

↑

Second Optimal Tableau (for $b_2 \downarrow$)

	Z	x_1	x_2	x_3	x_{s1}	x_{s2}	sol	BI
	0 :	3	1	3/2	0	1/2 :	3	x_2
	0 :	−11	0	−5	1	−2 :	0	x_{s1}
	1 :	21	0	7	0	4 :	24	Z

D sol	y_{s1}	y_{s2}	y_{s3}	y_1	y_2	R

The exploration of a further decrement of b_2 below the level of 6 units can now continue with the computation of another $-\Delta b_2$ such that

$$
\text{new } \mathbf{x}_B = \begin{bmatrix} 3 \\ 0 \end{bmatrix} + \begin{bmatrix} 1/2 \\ -2 \end{bmatrix} (-\Delta b_2) \geq 0
$$

Clearly, the second relation does not block Δb_2, since $2\Delta b_2 \geq 0$ and Δb_2 can be chosen arbitrarily large without violating the second inequality but, then, violating the first relation. Hence, from the first inequality

$$
-\Delta b_2 = (-3)/(1/2) = -6
$$

(Recall that with more than one inequality blocking the decrease of b_2, the decrement Δb_2 would have been computed as the minimum ratio among those blocking relations.) The new levels of the resource vector,

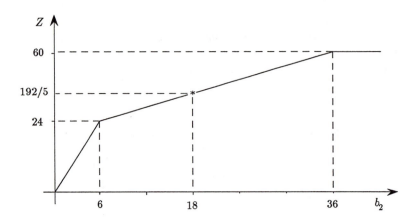

Figure 13.12. The revenue function with respect to input #2.

solution, and objective function are now

$$\text{new } \mathbf{b} \;=\; \begin{bmatrix} 12 \\ 6 \end{bmatrix} + \begin{bmatrix} 0 \\ -6 \end{bmatrix} \;=\; \begin{bmatrix} 12 \\ 0 \end{bmatrix}$$

$$\text{new } \mathbf{x}_B \;=\; \begin{bmatrix} 3 \\ 0 \end{bmatrix} + \begin{bmatrix} 1/2 \\ -2 \end{bmatrix}(-6) \;=\; \begin{bmatrix} x_2 = 0 \\ x_{s1} = 12 \end{bmatrix}$$

$$\text{new } Z \;=\; \begin{bmatrix} 8 & 0 \end{bmatrix}\begin{bmatrix} 0 \\ 12 \end{bmatrix} = 0$$

These results are consistent with the fact that the level of b_2 has been reduced to zero. No production of any commodity is possible because input #2 is a required resource for all outputs. The optimal solution exhibits the entire availability of input #1; $b_1 = 12$ as its only positive component. Since $b_2 = 0$, the parametric analysis of input #2 is complete. Although $b_2 = 0$, its shadow price is positive ($y_2 = 4$). This statement can be verified by attempting to change the last optimal tableau using the last solution as guideline. Recopying the last optimal tableau and replacing that solution with the new one corresponding to $b_2 = 0$, we have

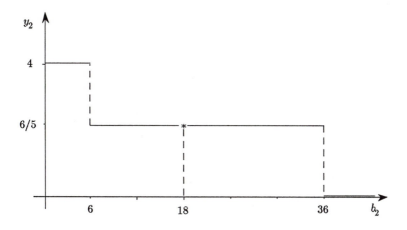

Figure 13.13. The inverse demand function for input #2.

Second Optimal Tableau (for $b_2 \downarrow$)

Z	x_1	x_2	x_3	x_{s1}	x_{s2}	sol	BI
0	3	1	3/2	0	1/2	0	$x_2 \rightarrow$
0	−11	0	−5	1	−2	12	x_{s1}
1	21	0	7	0	4	0	Z
D sol	y_{s1}	y_{s2}	y_{s3}	y_1	y_2	R	

In attempting to change the basis using the dual simplex algorithm, we notice that a pivot does not exist in the row corresponding to $x_2 = 0$ and the analysis stops. The entire parametric process for the derivation of the demand schedule of b_2 is summarized in the revenue and demand function diagrams of figures 13.12 and 13.13.

Chapter 14

Parametric Programming: Output Supply Functions

14.1 Derivation of Output Supply Functions

The post-optimality analysis of linear programming continues with the derivation of output supply functions. As in chapter 13, we briefly review the notion of output supply function and its derivation as presented in traditional economics courses and then draw the connection with a linear programming specification.

Given the production function $x = f(b_1, \ldots, b_m)$ where x is output and $b_1, \ldots, b_m$ are inputs, the profit maximization problem of a firm operating in a perfectly competitive market is stated as

$$\text{maximize } \pi = px - (r_1 b_1 + r_2 b_2 + \ldots + r_m b_m)$$

where p and $r_1, \ldots, r_m$ are output and input prices, respectively. First-order conditions for profit maximization are (section 13.2)

$$\frac{\partial \pi}{\partial b_i} = p \frac{\partial f(b_1, \ldots, b_m)}{\partial b_i} - r_i \overset{\text{set}}{=} 0, \qquad i = 1, \ldots, m$$

and represent the inverse demand functions for inputs. By solving this system of m equations, we obtain the direct demand functions that relate the quantity of the ith input to output and input prices:

$$b_i = g_i(p, r_1, \ldots, r_m), \qquad i = 1, \ldots, m$$

Finally, by substituting this optimal quantity b_i into the production function, the supply function for output x is obtained as

$$x = f[g_1(p, r_1, \ldots, r_m), \ldots, g_m(p, r_1, \ldots, r_m)] = h(p, r_1, \ldots, r_m)$$

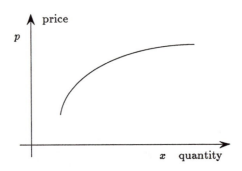

Figure 14.1. A traditional representation of the output supply function.

The output supply function, therefore, indicates the optimal quantity of commodity x to be produced, given the output and input prices. In order to represent the relation between the output quantity and the corresponding price p, the *ceteris paribus* principle is invoked once again. Keeping all the input prices constant and varying the output price p, the supply function is traced out as in figure 14.1.

Connection with Linear Programming. Analogous steps are at the foundation of supply functions in linear programming. Starting from an optimal tableau, which corresponds to the solution of the first-order conditions in the general approach,

1. Select a commodity whose supply function is to be developed; in general, we would like to derive supply functions for all commodities (outputs) in the model, one at a time.

2. Vary appropriately the price coefficient of the selected commodity while keeping all other prices constant (*ceteris paribus*).

3. Plot the results in two-dimensional diagrams.

Consider, then, the following numerical example:

Primal maximize $Z = 3x_1 + 8x_2 + 5x_3$

$$\text{subject to} \quad x_1 + 4x_2 + x_3 \leq 12$$
$$6x_1 + 2x_2 + 3x_3 \leq 18$$
$$x_j \geq 0, \quad j = 1, \ldots, 3$$

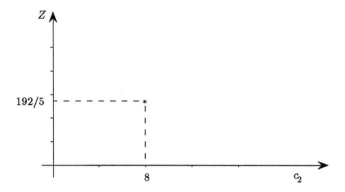

Figure 14.2. The revenue function for output #2.

The corresponding optimal tableau is

First Optimal Tableau

Z	x_1	x_2	x_3	x_{s1}	x_{s2}	P sol	BI
0	−3/10	1	0	3/10	−1/10	9/5	x_2
0	11/5	0	1	−1/5	2/5	24/5	x_3
1	28/5	0	0	7/5	6/5	192/5	Z
D sol	y_{s1}	y_{s2}	y_{s3}	y_1	y_2	R	

14.2 The Supply Function for Output #2

The choice of output #2 and the information of the optimal tableau allow the drawing of figures 14.2, 14.3, and 14.4, that is, the revenue function, the supply function, and the output space diagrams. Only one point can be plotted for the revenue and supply functions, justifying the need of a parametric programming approach.

The output space for the numerical example specified above requires a three-dimensional diagram, difficult to draw. The information of the optimal tableau, however, indicates that $x_1 = 0$ and, therefore, it is possible to draw a reduced output space diagram in the x_2 and x_3 coordinates without losing any essential information.

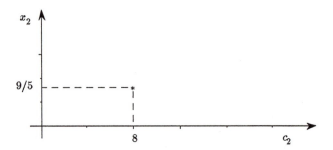

Figure 14.3. The supply function for output #2.

The first optimal tableau indicates that the extreme point A of figure 14.4 represents the optimal solution. At that point, the objective function has a slope entirely determined by the vector of price coefficients c, which is always perpendicular (normal, orthogonal) to the objective function line (plane in three or more dimensions; see section 5.5).

Suppose now we wish to increase the price of commodity #2, that is, the coefficient c_2 in the original objective function, and find out how the optimal solution changes. As we increase c_2 from 8 to a larger value, the objective function of figure 14.4 rotates clockwise toward constraint #1.

Correspondingly, the c vector (whose elements are the prices of commodities x_2 and x_3) rotates toward the new c vector. It is easy to verify the truth of these statements by simply noticing that the equation of the objective function passing through extreme point A is

$$x_3 = \frac{Z_A}{c_3} - \frac{c_2}{c_3}x_2$$

The slope of the objective function is given by $(-c_2/c_3)$ and, as c_2 increases (*ceteris paribus*), the slope becomes steeper. The vector c, which must always be 90° to the slope, rotates in the clockwise direction.

The relevant question can now be formulated: By how much can we increase c_2 without losing the optimality of extreme point A? In other equivalent words, By how much can we increase c_2 without the necessity of changing the current optimal basis? If we can find the critical value of c_2 requiring the changing of the optimal basis, we have found the relevant interval within which any value of c_2 corresponds to the same extreme point A. It should be clear, therefore, that we must worry only about the optimality criterion, since, (from figure 14.4) if we increase c_2 beyond a certain level the objective function will rotate at A, cutting through the feasible region, a definite indication of nonoptimality. The first step of

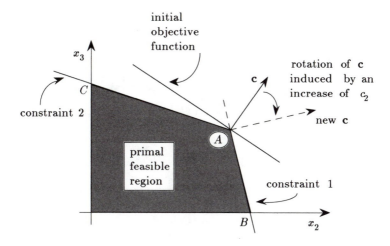

Figure 14.4. The output space (x_2, x_3).

the parametric analysis, therefore, requires increasing c_2 until the objective function becomes parallel to constraint #1. At that time, both extreme points A and B will represent basic optimal solutions.

14.2.1 The Opportunity Cost Revisited

The opportunity costs are directly affected by a change in any coefficient of the objective function. In the optimal tableau all the opportunity costs (associated with inequality constraints) are nonnegative. This is, indeed, the condition for determining that an optimal solution has been calculated. Finding the allowable variation of c_2, therefore, requires an analysis of the changes induced on the opportunity costs, which we wish to maintain at their nonnegative levels.

Before proceeding further, it is convenient to review the dual nature of the opportunity costs as discussed in section 7.3. The definition of opportunity cost is

$$(z_j - c_j) = \begin{cases} \mathbf{mrtt}_j' \mathbf{c}_B - c_j & \text{evaluated at market prices } (\textbf{primal}) \\ \mathbf{a}_j' \mathbf{y} - c_j & \text{evaluated at factor costs} \quad (\textbf{dual}) \end{cases}$$

To reinforce our confidence in this interesting pair of formulas, we compute the opportunity costs of commodity #1 using the information of the first optimal tableau. First, the primal version:

$$\begin{aligned}\text{opportunity cost of } x_1 &= (z_1 - c_1) = \mathbf{mrtt}_1' \mathbf{c}_B \quad - c_1\\ \text{at market prices}\end{aligned}$$

$$= [-3/10 \quad 11/5]\begin{bmatrix} 8 \\ 5 \end{bmatrix} - 3 = 28/5$$

The row vector $\mathbf{mrtt}_1' = [-3/10 \quad 11/5]$ is read as the column of the optimal tableau corresponding to activity #1. The vector

$$\mathbf{c}_B = \begin{bmatrix} 8 \\ 5 \end{bmatrix}$$

contains the coefficients of the objective function associated with the basic variables taken in order as they appear in the column of basic indexes.

The dual version of the opportunity cost is

$$\begin{aligned}\text{opportunity cost of } x_1 &= (z_1 - c_1) = \mathbf{a}_1' \mathbf{y} \quad - c_1\\ \text{at factor costs}\end{aligned}$$

$$= [1 \quad 6]\begin{bmatrix} 7/5 \\ 6/5 \end{bmatrix} - 3 = 28/5$$

The row vector $\mathbf{a}_1' = [1 \quad 6]$ represents the technology for producing one unit of commodity #1 and the vector

$$\mathbf{y} = \begin{bmatrix} 7/5 \\ 6/5 \end{bmatrix}$$

contains the imputed prices of the resources, or *factor costs*.

The same dual definition of opportunity cost applies to a commodity input. The primal version is

$$\begin{aligned}\text{opportunity cost of } x_{s1} &= (z_{s1} - c_{s1}) = \mathbf{mrtt}_{s1}' \mathbf{c}_B \quad - c_{s1} \quad = y_1\\ \text{at market prices}\end{aligned}$$

$$= [3/10 \quad -1/5]\begin{bmatrix} 8 \\ 5 \end{bmatrix} - 0 = 7/5$$

while the dual version is simply $(\mathbf{e}_1' = \mathbf{a}_{s1}')$

$$\begin{aligned}\text{opportunity cost of } x_{s1} &= (z_{s1} - c_{s1}) = \mathbf{e}_1' \mathbf{y} \quad - c_{s1} \quad = y_1\\ \text{at factor costs}\end{aligned}$$

$$= [1 \quad 0]\begin{bmatrix} 7/5 \\ 6/5 \end{bmatrix} - 0 = 7/5$$

From the above illustrations, we can derive an important generalization. By extending the notion of *opportunity cost at market prices* to all inputs at once, we can write

$$\mathbf{y} = [MRTT_s'] \mathbf{c}_B$$

where, as usual, c_B is the vector of objective function coefficients corresponding to basic variables and $[MRTT_s]$ is the matrix of marginal rates of technical transformation corresponding to all inputs. Therefore, the real connection between opportunity costs evaluated at factor cost and at market prices can be revealed by replacing the vector of input shadow prices $\mathbf{y}$ in the dual definition

$$
\begin{aligned}
(z_j - c_j) &= \mathbf{a}_j'\mathbf{y} & -\ c_j & \quad \text{dual definition} \\
&= \mathbf{a}_j'[MRTT_s']\mathbf{c}_B & -\ c_j & \quad \mathbf{y} \text{ connection} \\
&= \mathbf{mrtt}_j'\mathbf{c}_B & -\ c_j & \quad \text{primal definition}
\end{aligned}
$$

We conclude that $\mathbf{mrtt}_j' = \mathbf{a}_j'[MRTT_s']$; that is, the marginal rates of technical transformation of any commodity output can be obtained as the product of the original technical coefficients for producing that output and the marginal rates of technical substitution between the inputs involved in the production plan.

A change in the vector of revenue coefficients $\mathbf{c}$ and its consequences on the dual solution can now be stated in general form by using all the accumulated knowledge. A change in the $\mathbf{c}$ vector can be indicated as

$$
\text{new } \mathbf{c} = \text{old } \mathbf{c} + \Delta\mathbf{c} = \mathbf{c} + \Delta\mathbf{c}
$$

which is analogous to the change in the $\mathbf{b}$ vector discussed in chapter 13. A change in the vector of dual variables $\mathbf{y}$ can similarly be stated as

$$
\text{new } \mathbf{y} = \text{old } \mathbf{y} + \Delta\mathbf{y} = \mathbf{y} + \Delta\mathbf{y}
$$

but we have just seen that $\mathbf{y} = [MRTT_s']\mathbf{c}_B$ and, therefore,

$$
\Delta\mathbf{y} = [MRTT_s']\Delta\mathbf{c}_B
$$

Let us restate at this point the objective of the analysis. We wish to trace out the effects of a changing $\mathbf{c}$ vector while maintaining the feasible dual solution (which is equivalent to maintaining the optimal primal solution). With a new $\mathbf{c}$ we naturally have new $(z_j - c_j)$. Hence, by the dual definition of opportunity cost we can infer the following chain of deductions:

$$
\begin{aligned}
\text{new } (z_j - c_j) &= \mathbf{a}_j'(\text{new } \mathbf{y}) & - (\text{new } c_j) & \quad \geq 0 \\
&= \mathbf{a}_j'(\mathbf{y} + \Delta\mathbf{y}) - (c_j + \Delta c_j) & & \quad \geq 0 \quad \text{substituting} \\
&= (\mathbf{a}_j'\mathbf{y} - c_j) \ + \mathbf{a}_j'\Delta\mathbf{y} - \Delta c_j) & & \quad \geq 0 \quad \text{rearranging} \\
&= \text{old}(z_j - c_j) + (\mathbf{a}_j'[MRTT_s']\Delta\mathbf{c}_B - \Delta c_j) \geq 0
\end{aligned}
$$

If the activity whose revenue coefficient is being changed is in the optimal basis, then

$$\Delta \mathbf{c}_B = \begin{bmatrix} 0 \\ \cdots \\ \Delta c_j \\ \cdots \\ 0 \end{bmatrix} = \begin{bmatrix} 0 \\ \cdots \\ 1 \\ \cdots \\ 0 \end{bmatrix} \Delta c_j$$

and, furthermore, notice that the elements of the unit vector can be represented using the Kronecker δ-notation as

$$\delta_{kj} = \begin{cases} 1 & \text{if } k = j \\ 0 & \text{if } k \neq j \end{cases} \qquad \text{for } k, j = 1, \dots, m$$

With this notation, the optimality criterion for the new scenario can be rewritten as

$$\text{new } (z_j - c_j) = \text{old } (z_j - c_j) + (\mathbf{a}'_j \mathbf{mrtt}'_{sk} - \delta_{kj})\Delta c_j \geq 0$$

from which we can compute the maximum allowable increment of c_j as

$$\text{old } (z_j - c_j) \geq -(\mathbf{a}'_j \mathbf{mrtt}'_{sk} - \delta_{kj})\Delta c_j$$

$$\Delta c_j \leq \frac{\text{old } (z_j - c_j)}{-(\mathbf{a}'_j \mathbf{mrtt}'_{sk} - \delta_{kj})}$$

and, finally,

$$\Delta c_j = \min_j \left\{ \frac{\text{old } (z_j - c_j)}{-(\mathbf{a}'_j \mathbf{mrtt}'_{sk} - \delta_{kj})}, \quad \text{where} \quad -(\mathbf{a}'_j \mathbf{mrtt}'_{sk} - \delta_{kj}) > 0 \right\}$$

In words, the admissible increment of c_j is that value of Δc_j computed as the minimum of all possible ratios between the components of the old dual solution and the corresponding positive denominators. The minimum ratio is selected to preserve the feasibility of the new dual solution or, equivalently, to preserve the optimality of the primal solution.

Before implementing the results of the above discussion, we must know how to compute the coefficient $(\mathbf{a}'_j \mathbf{mrtt}'_{sk})$ in the denominator of the minimum ratio. This operation is simple because that quantity is already computed and appears in the first optimal tableau. In fact, since we are considering a parametric variation in the price of output #2, x_2, that index appears as the first component of the basic optimal solution. The $\mathbf{mrtt}'_{s2}$ vector associated with x_2 is $[3/10 \quad -1/10]$, the vector under the slack variables in the first row of the optimal tableau (because output x_2 was chosen for parametric analysis).

To verify that the desired quantity $(\mathbf{a}'_j \mathbf{mrtt}'_{s2})$ is indeed the jth coefficient in the first row (the row corresponding to the position of the selected variable in the basic solution), we compute the five coefficients as

$$(\mathbf{a}'_1 \mathbf{mrtt}'_{s2}) = [1 \quad 6] \begin{bmatrix} 3/10 \\ -1/10 \end{bmatrix} = \frac{3}{10} - \frac{6}{10} = -\frac{3}{10}$$

$$(\mathbf{a}'_2 \mathbf{mrtt}'_{s2}) = [4 \quad 2] \begin{bmatrix} 3/10 \\ -1/10 \end{bmatrix} = \frac{12}{10} - \frac{2}{10} = 1$$

$$(\mathbf{a}'_3 \mathbf{mrtt}'_{s2}) = [1 \quad 3] \begin{bmatrix} 3/10 \\ -1/10 \end{bmatrix} = \frac{3}{10} - \frac{3}{10} = 0$$

$$(\mathbf{e}'_1 \mathbf{mrtt}'_{s2}) = [1 \quad 0] \begin{bmatrix} 3/10 \\ -1/10 \end{bmatrix} = \frac{3}{10}$$

$$(\mathbf{e}'_2 \mathbf{mrtt}'_{s2}) = [0 \quad 1] \begin{bmatrix} 3/10 \\ -1/10 \end{bmatrix} = -\frac{1}{10}$$

The above five coefficients are indeed the same coefficients appearing in the first row of the optimal tableau.

14.2.2 Increase in Price c_2 (the Stolper-Samuelson Theorem)

We now return to our original purpose, which is that of analyzing the effects of a change in the revenue coefficient c_2 while keeping all other parameters constant. We must now determine the allowable increment Δc_2 using the optimality criterion developed in the previous section. Hence,

$$\text{new } (z_j - c_j) = \text{old } (z_j - c_j) + (\mathbf{a}'_j \mathbf{mrtt}'_{s2} - \delta_{2j}) \Delta c_2 \geq 0$$

$$= \begin{bmatrix} 28/5 \\ 0 \\ 0 \\ 7/5 \\ 6/5 \end{bmatrix} + \left\{ \begin{bmatrix} -3/10 \\ 1 \\ 0 \\ 3/10 \\ -1/10 \end{bmatrix} - \begin{bmatrix} 0 \\ 1 \\ 0 \\ 0 \\ 0 \end{bmatrix} \right\} \Delta c_2 \geq 0$$

The last vector in the braces represents the values of δ_{2j} with $\delta_{22} = 1$ and $\delta_{2j} = 0$ for $j \neq 2$, as stated above. The admissible increment Δc_2 is computed according to the minimum ratio developed in the previous section, that is,

$$\Delta c_2 = \min \left\{ \frac{28/5}{-(-3/10)}, \frac{6/5}{-(-1/10)} \right\} = \left\{ \frac{56}{3}, \frac{12}{1} \right\} = 12$$

Thus, the admissible new value of c_2 is

$$\text{new } c_2 = \text{old } c_2 + \Delta c_2 = 8 + 12 = 20$$

All other prices remain at the initial value. The new row of opportunity costs is computed by substituting the value of $\Delta c_2 = 12$ into the new $(z_j - c_j)$ coefficients as

$$\text{new } (z_j - c_j) = \begin{bmatrix} 28/5 \\ 0 \\ 0 \\ 7/5 \\ 6/5 \end{bmatrix} + \left\{ \begin{bmatrix} -3/10 \\ 1 \\ 0 \\ 3/10 \\ -1/10 \end{bmatrix} - \begin{bmatrix} 0 \\ 1 \\ 0 \\ 0 \\ 0 \end{bmatrix} \right\} 12 = \begin{bmatrix} 2 \\ 0 \\ 0 \\ 5 \\ 0^* \end{bmatrix}$$

The starred zero indicates that the new dual solution is degenerate, as anticipated. The new value of the objective function is

$$Z = \begin{bmatrix} 20 & 5 \end{bmatrix} \begin{bmatrix} 9/5 \\ 24/5 \end{bmatrix} = 36 + 24 = 60$$

The new $(z_j - c_j)$ vector contains a degenerate component corresponding to the element that determined the maximum allowable increment Δc_2. Since the basis has not changed, this is an indication of dual degeneracy that should allow the computation of an alternative optimal solution (extreme point B) as illustrated in the output space of figure 14.4.

In order to continue exploring the increase in c_2, we must change the basis to preserve the optimality of the solution. We transcribe, therefore, the optimal tableau, replacing the old $(z_j - c_j)$ row with the new one, and compute the new optimal solution using the primal simplex algorithm.

First Optimal Tableau ($c_2 \uparrow$)

		Z		x_1	x_2	x_3	x_{s1}	x_{s2}		sol	BI
1	1/4 : 0		0 :	−3/10	1	0	3/10	−1/10	:	9/5	x_2
0	5/2 : 0		0 :	11/5	0	1	−1/5	$\boxed{2/5}$	:	24/5	x_3 →
0	0 : 1		1 :	2	0	0	5	0*	:	60	Z

dual degeneracy ↑

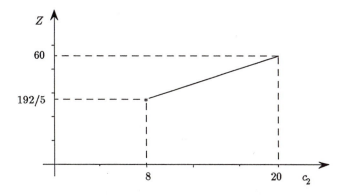

Figure 14.5. The revenue function at extreme point B.

Second Optimal Tableau $(c_2 \uparrow)$

Z	x_1	x_2	x_3	x_{s1}	x_{s2}	sol	BI
0	1/4	1	1/4	1/4	0	3	x_2
0	11/2	0	5/2	−1/2	1	12	x_{s2}
1	2	0	0	5	0	60	Z

The solution in the second optimal tableau corresponds to extreme point B in figure 14.4. The justification for the change of basis from extreme point A to B represented in the above computations is based upon the fact that an increase of 12 in the price of activity #2 (originally set at 8) has aligned the objective function Z with constraint #1. A further increase (above 20) of c_2 without changing the basis corresponding to extreme point A would rotate the objective function in such a way as to cut through the feasible region thus losing the optimality of the solution. Before the occurrence of this event, parametric programming suggests changing the optimal basis from extreme point A to B, a process described in the two optimal tableaux.

With the information developed up to this point, the revenue and supply function diagrams can be extended from their original dotlike representation in figures 14.2 and 14.3 to assume the form of figures 14.5 and 14.6.

The Stolper-Samuelson theorem. Dual to the Rybczynski theorem,

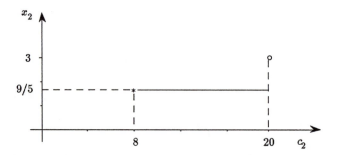

Figure 14.6. The supply function for x_2 at extreme point B.

which states response relations between inputs and outputs (physical quantities), a proposition exists known as *the Stolper-Samuelson theorem* that establishes definite relations among output and input prices. This theorem says that (given outputs using the same inputs) when the price of the more input-intensive output increases, the price of that input will increase more than proportionately to the increase of the output price. Conversely, the price of the other input will decrease more than proportionately.

In the numerical example we are discussing, activity a_2 is more input #1–intensive than activity a_3, as already indicated in section 13.4.1. At the original equilibrium, with price of output #2 equal to 8 and of output #3 equal to 5, the input shadow prices are those indicated in the first optimal tableau, namely, $y_1 = 7/5$ and $y_2 = 6/5$. The increase in c_2 (the price of the input #1–intensive activity) from 8 to 20 causes an increase of the shadow price of input #1, y_1, from 7/5 to 5, as indicated by the second optimal tableau. Conversely, the price of the other input, y_2, drops to zero from an original equilibrium level of 6/5.

Figure 14.7 illustrates these output-input price relationships by means of the dual input-price space. The graph of this space is obtained from the dual system of constraints involving the prices of inputs #1 and #2 and the prices of output # 2 and #3 (commodity #1 is at the zero level):

$$4y_1 + 2y_2 = 8 = c_2$$
$$y_1 + 3y_2 = 5 = c_3$$

The right shift of the dual constraint #2 corresponds to an increase of c_2, which causes an increase in the shadow price of input #1, y_1, and a decrease of y_2, as predicted by the Stolper-Samuelson theorem.

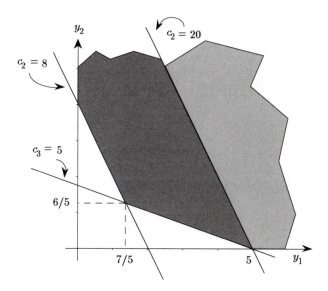

Figure 14.7. The Stolper-Samuelson theorem illustrated.

14.2.3   Further Increase of c_2

An increase of c_2 beyond the level of 20 requires the computation of a new row of opportunity costs $(z_j - c_j)$ using the second optimal tableau corresponding to extreme point B. The relevant diagram of the output space is now represented in figure 14.8. By repeating the reasoning developed in previous sections, it is possible to increase c_2 beyond the level of 20, thus causing the rotation of the objective function around extreme point B without losing the optimality of the current solution.

The admissible new increment of c_2 is determined through a new set of $(z_j - c_j)$ using the information of the last optimal tableau as follows:

$$\text{new } (z_j - c_j) = \text{old } (z_j - c_j) + (\text{first row} - \delta_{2j})\Delta c_2 \geq 0$$

$$= \begin{bmatrix} 2 \\ 0 \\ 0 \\ 5 \\ 0 \end{bmatrix} + \left\{ \begin{bmatrix} 1/4 \\ 1 \\ 1/4 \\ 1/4 \\ 0 \end{bmatrix} - \begin{bmatrix} 0 \\ 1 \\ 0 \\ 0 \\ 0 \end{bmatrix} \right\} \Delta c_2 \geq 0$$

We immediately notice that there are no negative elements in the braces, as required by the blocking criterion for Δc_2 and expressed by

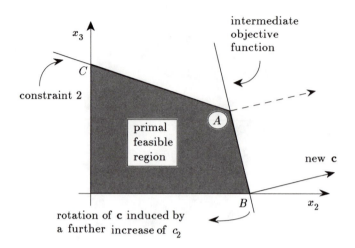

Figure 14.8. Output space with the optimal solution at extreme point B.

the minimum ratio of section 14.2.1. The admissible increment Δc_2, in other words, can be chosen as large as $+\infty$ without violating the above inequalities and without losing the optimality of the current solution. Hence, the parametric analysis in the direction of a price increase for activity #2 is completed. The revenue and supply function diagrams can be further extended as in figures 14.9 and 14.10.

The slope of the objective function beyond the price level of $c_2 = 20$ corresponds to the derivative of the objective function with respect to the price of commodity #2. This type of derivative was discussed in section 1.10. In this example, the objective function has the following representation:

$$Z^* = c_1 x_1^* + c_2 x_2^* + c_3 x_3^*$$

where the asterisk indicates optimal values. The relevant derivative, defined in the interval $(20, +\infty)$,

$$\frac{\partial Z^*}{\partial c_2} = x_2^* = 3$$

gives the slope of the last segment of the revenue function in figure 14.9.

14.2.4 Decrease in Price c_2

The procedure developed for a price increase can readily be adapted for a price decrease. For convenience, we define a price decrease as $-\Delta c_2$. Hence, starting from the first optimal tableau, we compute the maximum admissi-

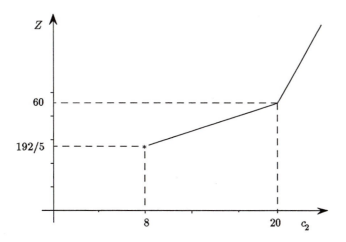

Figure 14.9. The revenue function for the increase of c_2.

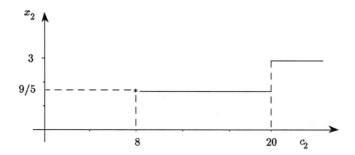

Figure 14.10. The supply function for x_2 for the increase of c_2.

ble decrement of c_2 (from the level of 8) using the new $(z_j - c_j)$ criterion:

$$\text{new } (z_j - c_j) = \text{old } (z_j - c_j) + \quad (\text{first row} \quad - \quad \delta_{2j})(-\Delta c_2) \quad \geq 0$$

$$= \begin{bmatrix} 28/5 \\ 0 \\ 0 \\ 7/5 \\ 6/5 \end{bmatrix} + \left\{ \begin{bmatrix} -3/10 \\ 1 \\ 0 \\ 3/10 \\ -1/10 \end{bmatrix} - \begin{bmatrix} 0 \\ 1 \\ 0 \\ 0 \\ 0 \end{bmatrix} \right\} (-\Delta c_2) \geq 0$$

Taking into account the $(-)$ sign in front of Δc_2, the blocking criterion expressed by the minimum ratio gives the following result:

$$\Delta c_2 = \min \left\{ \frac{7/5}{-(-3/10)} \right\} = \frac{14}{3}$$

Thus, a decrease of 14/3 (from 8) in the price of activity #2 can be realized without changing the optimal basis $[\mathbf{a}_2, \mathbf{a}_3]$. The new row of opportunity costs corresponding to the reduction of c_2 from 8 to 10/3 is

$$\text{new } (z_j - c_j) = \begin{bmatrix} 7 \\ 0 \\ 0 \\ 0^* \\ 5/3 \end{bmatrix} \leftarrow \text{dual degeneracy}$$

The new c_2 is equal to $8 - 14/3 = 10/3$ and the value of the objective function is

$$Z = [10/3 \quad 5] \begin{bmatrix} 9/5 \\ 24/5 \end{bmatrix} = 30$$

Before continuing the analysis of a decrease in c_2, the basis must be changed. This is done, as before, by transcribing the first optimal tableau, replacing the old $(z_j - c_j)$ with the new values, and pivoting on the activity corresponding to the degenerate value of the new dual solution.

First Optimal Tableau ($c_2 \downarrow$)

$$\begin{bmatrix} 10/3 & 0 & \vdots & 0 \\ 2/3 & 1 & \vdots & 0 \\ \cdots & \cdots & & \cdots \\ 0 & 0 & \vdots & 1 \end{bmatrix}$$

Z	x_1	x_2	x_3	x_{s1}	x_{s2}	sol	BI
0 :	−3/10	1	0	$\boxed{3/10}$	−1/10 :	9/5	$x_2 \rightarrow$
0 :	11/5	0	1	−1/5	2/5 :	24/5	x_3
1 :	7	0	0	0*	5/3 :	30	Z

dual degeneracy ↑

Second Optimal Tableau ($c_2 \downarrow$)

Z	x_1	x_2	x_3	x_{s1}	x_{s2}	sol	BI
0 :	−1	10/3	0	1	−1/3 :	6	x_{s1}
0 :	2	2/3	1	0	1/3 :	6	x_3
1 :	7	0	0	0	5/3 :	30	Z

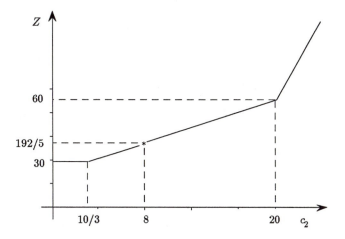

Figure 14.11. The complete revenue function for c_2.

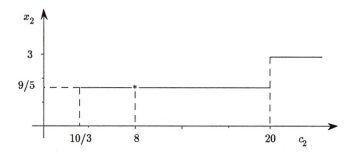

Figure 14.12. The complete supply function for x_2.

Activity #2 has left the optimal basis and, therefore, $x_2 = 0$. Any further decrease of c_2 would obviously maintain the unprofitability of activity #2 and, as a consequence, the parametric analysis of c_2 is terminated. The revenue and supply functions diagrams can now be completed as in figures 14.11 and 14.12.

The first step (increase and decrease of c_2) in the supply function diagram is called *sensitivity analysis*. With this terminology we wish to describe the process of testing how sensitive the original optimal basis is to perturbations of the various parameters (in this case price c_2). It can also be viewed as the methodology for measuring the original equilibrium displacement, that is, by how much the original optimal solution (dual solution, in this case) will vary in relation to changes of the various parameters.

14.3 The Supply Function for Output #3

In order to review and reinforce the process of conducting a parametric analysis of the price coefficients, we derive the supply and revenue functions for commodity #3. We will avoid a lengthy and largely repetitive discussion and simply stylize the various steps leading to a complete analysis.

14.3.1 Increase of c_3

Using the information in the first optimal tableau, we must determine the admissible increment Δc_3 via the new $(z_j - c_j)$ criterion:

$$\text{new } (z_j - c_j) = \text{old } (z_j - c_j) + (\mathbf{a}'_j \mathbf{mrtt}'_{s3} - \delta_{3j})\Delta c_3 \geq 0$$

$$= \begin{bmatrix} 28/5 \\ 0 \\ 0 \\ 7/5 \\ 6/5 \end{bmatrix} + \left\{ \begin{bmatrix} 11/5 \\ 0 \\ 1 \\ -1/5 \\ 2/5 \end{bmatrix} - \begin{bmatrix} 0 \\ 0 \\ 1 \\ 0 \\ 0 \end{bmatrix} \right\} \Delta c_3 \geq 0$$

Solving for Δc_3 (we indicate the minimum criterion as a reminder of the process even though there is only one element to choose from),

$$\Delta c_3 = \min \left\{ \frac{7/5}{-(-1/5)} \right\} = 7$$

Hence, new $c_3 = $ old $c_3 + \Delta c_3 = 5 + 7 = 12$; the new opportunity costs are

$$\text{new } (z_j - c_j) = \begin{bmatrix} 21 \\ 0 \\ 0 \\ 0* \\ 4 \end{bmatrix} \leftarrow \text{dual degeneracy}$$

The new value of the objective function is

$$Z = \begin{bmatrix} 8 & 12 \end{bmatrix} \begin{bmatrix} 9/5 \\ 24/5 \end{bmatrix} = 72$$

A change of basis with $c_3 = 12$ is required. Use the primal simplex algorithm.

First Optimal Tableau ($c_3 \uparrow$)

$$
\begin{bmatrix}
10/3 & 0 & \vdots & 0 \\
2/3 & 1 & \vdots & 0 \\
\hdotsfor{4} \\
0 & 0 & \vdots & 1
\end{bmatrix}
$$

	Z	x_1	x_2	x_3	x_{s1}	x_{s2}	sol	BI
	0	$-3/10$	1	0	$\boxed{3/10}$	$-1/10$	9/5	$x_2 \rightarrow$
	0	11/5	0	1	$-1/5$	2/5	24/5	x_3
	1	21	0	0	0^*	4	72	Z

dual degeneracy $\uparrow$

Second Optimal Tableau ($c_3 \uparrow$)

Z	x_1	x_2	x_3	x_{s1}	x_{s2}	sol	BI
0	-1	10/3	0	1	$-1/3$	6	x_{s1}
0	2	2/3	1	0	1/3	6	x_3
1	21	0	0	0	4	72	Z

A further increase in c_3 can be effected by repeating the process based on the information of the last optimal tableau:

$$
\text{new } (z_j - c_j) = \text{old } (z_j - c_j) + (\text{second row} - \delta_{3j})\Delta c_3 \geq 0
$$

$$
= \begin{bmatrix} 21 \\ 0 \\ 0 \\ 0 \\ 4 \end{bmatrix} + \left\{ \begin{bmatrix} 2 \\ 2/3 \\ 1 \\ 0 \\ 1/3 \end{bmatrix} - \begin{bmatrix} 0 \\ 0 \\ 1 \\ 0 \\ 0 \end{bmatrix} \right\} \Delta c_3 \geq 0
$$

Since all the elements in the braces are nonnegative, Δc_3 can increase indefinitely without causing the infeasibility of the dual solution or, equivalently, without loss of optimality of the primal solution. The analysis for an increase of c_3 is complete.

14.3.2 Decrease in c_3

From the first optimal tableau and using $-\Delta c_3$ to indicate the reduction of c_3, we compute the new opportunity costs:

$$\text{new } (z_j - c_j) = \text{old } (z_j - c_j) + (\text{second row} - \delta_{3j})(-\Delta c_3) \geq 0$$

$$= \begin{bmatrix} 28/5 \\ 0 \\ 0 \\ 7/5 \\ 6/5 \end{bmatrix} + \left\{ \begin{bmatrix} 11/5 \\ 0 \\ 1 \\ -1/5 \\ 2/5 \end{bmatrix} - \begin{bmatrix} 0 \\ 0 \\ 1 \\ 0 \\ 0 \end{bmatrix} \right\} (-\Delta c_3) \geq 0$$

Hence,

$$\Delta c_3 = \min \left\{ \frac{28/5}{-(-11/5)}, \quad \frac{6/5}{-(-2/5)} \right\} = \left\{ \frac{28}{11}, \frac{3}{1} \right\} = \frac{28}{11}$$

The new price level is new $c_3 = 5 - 28/11 = 27/11$. The new opportunity costs are

$$\text{new } (z_j - c_j) = \begin{bmatrix} 0^* \\ 0 \\ 0 \\ 21/11 \\ 2/11 \end{bmatrix} \leftarrow \text{dual degeneracy}$$

The new value of the objective function is

$$Z = \begin{bmatrix} 8 & 27/11 \end{bmatrix} \begin{bmatrix} 9/5 \\ 24/5 \end{bmatrix} = \frac{288}{11}$$

We need to change basis for a further decrease of c_3.

First Optimal Tableau ($c_3 \downarrow$)

$$\begin{bmatrix} 1 & 3/22 & \vdots & 0 \\ 0 & 5/11 & \vdots & 0 \\ \cdots & \cdots \\ 0 & 0 & \vdots & 1 \end{bmatrix} \begin{bmatrix} & Z & x_1 & x_2 & x_3 & x_{s1} & x_{s2} & & \text{sol} & \text{BI} \\ 0 & \vdots & -3/10 & 1 & 0 & 3/10 & -1/10 & \vdots & 9/5 & x_2 \\ 0 & \vdots & \boxed{11/5} & 0 & 1 & -1/5 & 2/5 & \vdots & 24/5 & x_3 \rightarrow \\ \cdots & \cdots \\ 1 & \vdots & 0^* & 0 & 0 & 21/11 & 2/11 & \vdots & 288/11 & Z \end{bmatrix}$$

$\uparrow$ dual degeneracy

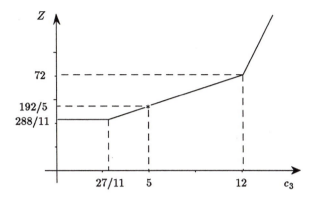

Figure 14.13. The revenue function for c_3.

Second Optimal Tableau ($c_3 \downarrow$)

$$
\begin{array}{cccccccc}
Z & x_1 & x_2 & x_3 & x_{s1} & x_{s2} & \text{sol} & \text{BI} \\
\left[\begin{array}{c} 0 \\ 0 \\ \\ 1 \end{array}\right.
& \begin{array}{c} 0 \\ 1 \\ \\ 0 \end{array}
& \begin{array}{c} 1 \\ 0 \\ \\ 0 \end{array}
& \begin{array}{c} 3/22 \\ 5/11 \\ \\ 0 \end{array}
& \begin{array}{c} 3/11 \\ -1/11 \\ \\ 21/11 \end{array}
& \begin{array}{c} -1/22 \\ 2/11 \\ \\ 2/11 \end{array}
& \left.\begin{array}{c} 27/11 \\ 24/11 \\ \\ 288/11 \end{array}\right]
& \begin{array}{c} x_2 \\ x_1 \\ \\ Z \end{array}
\end{array}
$$

The analysis is concluded because activity #3 has left the basis and, therefore, $x_3 = 0$ with a price level of $c_3 = 27/11$. With a lower price activity #3 would be even more unprofitable. The parametric information concerning the supply function of output #3 is summarized in the revenue and supply diagrams of figures 14.13 and 14.14.

14.4 Parametric Analysis of a Price Whose Activity Is Not in the Optimal Basis

The price variation for an activity not appearing in the optimal basis is very simple and direct. It does not involve the process developed above. Consider the price of output #1, c_1. The information for determining by how much c_1 should change before activity #1 is brought into the basis is contained in the first optimal tableau of section 14.1. In fact, since the loss is positive, $(z_1 - c_1) = 28/5$, it is obvious that activity #1 is not profitable.

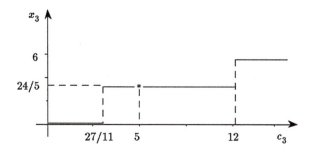

Figure 14.14. The supply function for x_3.

In this case, decreasing c_1 will not make it any more convenient. We need to consider, therefore, only by how much c_1 should increase before the corresponding activity breaks even and becomes a candidate for entering the production plan (basis). The answer is

$$z_1 = c_1 + \frac{28}{5} = 3 + \frac{28}{5} = \frac{43}{5}$$

or, in words, c_1 must increase at least as much as the present loss of 28/5. At that point, activity #1 is brought into the basis and the parametric procedure developed in this chapter applies in full.

The information developed by parametric programming also involves the cross-variations between parameters and variables. For example, with the additional information contained in the various optimal tableaux computed in previous sections, it would be possible to draw diagrams relating the variation of the price of activity #2 with the output quantity of activity #3.

14.5 Symmetry in Parametric Programming

It is inevitable to conclude the discussion of parametric programming by drawing attention to the symmetric relations developed in the derivation of input demand and output supply functions. The simplest way to exhibit this symmetry is to collect four figures (as in figure 14.15), two from the input demand (figures 13.10 and 13.11) and two from the output supply analyses (figures 14.11 and 14.12). This symmetry reveals that

1. The revenue function is concave with respect to limiting inputs.
2. The revenue function is convex with respect to output prices.
3. The inverse demand function is nonincreasing.

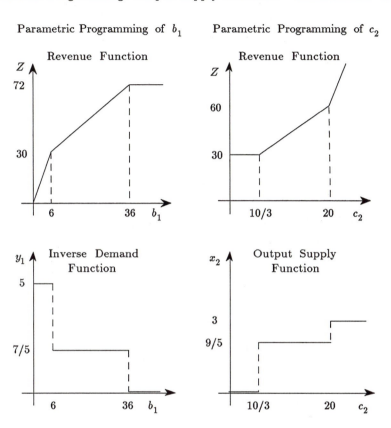

Figure 14.15. The symmetry of parametric programming.

4. The output supply function is nondecreasing.

The form of these functions is consistent with economic theory and its requirements for representing rational behavior of economic agents.

Chapter 15

Dealing with Multiple Optimal Solutions

15.1 Introduction

For many years, LP users have regarded multiple optimal solutions either as an exceptional event or as a nuisance to be avoided. Indeed, in many circles, multiple optimal solutions are a source of embarrassment and, often, the main goal of researchers is to define sufficient conditions for unique solutions.

We hold the opposite point of view. In any relevant problem, multiple optimal solutions are the rule, not the exception. Furthermore, the existence of multiple optimal solutions is an indication that the problem is rich in technical and economic information. The existence of multiple optimal solutions makes the final selection strategy a real problem of choice to be determined with criteria other than mathematical programming. On the contrary, problems with unique solutions are suspect as to their relevance. Their existence at the macro level provides a logical basis for a dictatorial view of government. Hence, when a linear programming problem exhibits multiple optimal solutions (whether primal or dual), it means that the problem at hand is potentially more relevant than a similar problem that exhibits unique optimal solutions.

If the firms analyzed by linear programming are assumed to operate under perfectly competitive conditions, and if the LP model is really capable of reflecting such an environment, one would expect many activities to be equally profitable, whether or not they are included in the optimal basis. This is because zero profitability is a long-run necessary condition for competitive markets. In general, commodities are available on these markets in larger number than the linear programming constraints of a single firm.

Let us reformulate the idea as follows: in a competitive situation, where there are many firms using essentially the same technology, one observes that similar firms produce different product mixes. That is, all activities are equally profitable and, therefore, it just happens that one firm chooses a particular combination of activities while others select a different mix. In LP terminology, this situation is characterized by zero loss not only for the basic activities but also for those not in the basis. It causes the multiple optimal solutions phenomenon. Hence, an extensive dual degeneracy may be interpreted as a validation of a LP economic model where perfect competition prevails.

In section 8.5 we found that in order to have multiple optimal solutions a problem must first exhibit degeneracy. Degeneracy of the dual solution is a necessary condition for the existence of primal multiple optimal solutions. Degeneracy of the dual solution means that an activity not in the primal basis, say the jth activity, exhibits a break-even property according to which the corresponding opportunity cost is equal to zero; in symbolic terms $(z_j - c_j) = 0$. Degeneracy of the primal solution means that the level of a basic primal activity is equal to zero and it is a necessary condition for the existence of dual optimal solutions. In both cases, the sufficient conditions for multiple optimal solutions are the degeneracy and the possibility of finding a pivot.

Degeneracy of a solution is often viewed with little favor by numerical analysts because it causes ties in the simplex algorithms when minimum ratios are computed for moving from one extreme point to an adjacent one. In principle, this fact could cause cycling, which corresponds to an infinite loop for the algorithm (a path departing from an extreme point returns to the same starting point and the cycle continues). In practice, however, cycling is an extremely rare phenomenon encountered only with very peculiar linear programming structures. Based on the experience accumulated to date, it need not worry the economic researcher.

To avoid degeneracy, the suggestion coming from various quarters is to slightly perturb the solution, therefore avoiding any tie and possible cycling. In this way the possibility of multiple optimal solutions is also eliminated. From our perspective, it seems more interesting to take a very small risk for the cycling of the simplex algorithm while seeking the various multiple optimal solutions. In fact we recommend using perturbation to achieve degeneracy of the optimal solution rather than to avoid it.

It may be useful to restate here the reason why multiple optimal solutions are important for empirical analysis. As we observe farmers making their enterprise selection, we notice the great variety of crops cultivated by them, especially in California. On the contrary, the limiting resources are relatively few: land qualities, water, operating and investment capital, specialized labor, and a few others. How is it conceivable that linear programming constitutes a reasonable approach for representing such a diversified mix of crop and animal activities in the presence of a relatively few constraints? The question is legitimate because we have learned that a basic solution cannot have more positive components than constraints. If it were not for the possibility of multiple optimal solutions, it would be impossible to propose linear programming as a framework for representing and analyzing the richness of farmers' decisions. Hence, while nondegenerate linear programming problems imply specialization of a production plan, degenerate LP offers the possibility for diversified activity mixes, just as we observe in real life.

To illustrate the issues and the associated answers, consider the example in the following section.

15.2   Choosing among Multiple Optimal Solutions

Last year, a vintner produced sparkling wine, red and white table wine, and made his own bottles. The following quantities were recorded (forgive the fractional bottles):

sparkling wine	observed x_1	=	4/9
red table wine	observed x_2	=	2/17
bottles	observed x_3	=	329/621
white table wine	observed x_4	=	1/11

Unsure whether his operation is as efficient as he would like it to be, he hires a linear programming consultant and asks her to evaluate his winemaking operation. The entrepreneur makes available to the consultant all the technical and financial information she requests. After a few days she is able to set up the following linear programming problem:

Primal maximize $Z = \dfrac{53}{22}x_1 + \dfrac{39}{22}x_2 + 5x_3 + 2x_4$

$$\text{subject to} \quad
\begin{array}{rcrcrcrcl}
3x_1 & + & 2x_2 & + & x_3 & + & 4x_4 & \leq & 6 \\
2x_1 & + & x_2 & + & 5x_3 & + & x_4 & \leq & 4 \\
x_1 & + & 3x_2 & - & 2x_3 & + & 4x_4 & \leq & 0
\end{array}$$

all $x_j \geq 0$

This linear programming problem exhibits multiple primal optimal solutions as the following tableaux indicate:

First Optimal Tableau

Z	x_1	x_2	x_3	x_4	x_{s1}	x_{s2}	x_{s3}	sol	BI
0 :	23/22	−25/22	0	0	1	−6/11	−19/22	: 42/11	x_{s1}
0 :	7/22	1/22	1	0	0	2/11	−1/22	: 8/11	x_3
0 :	9/22	$\boxed{17/22}$	0	1	0	1/11	5/22	: 4/11	x_4 →
1 :	0	0	0	0	0	12/11	5/22	: 48/11	Z

(↑ under x_2)

Second Optimal Tableau

Z	x_1	x_2	x_3	x_4	x_{s1}	x_{s2}	x_{s3}	sol	BI
0 :	28/17	0	0	25/17	1	−7/17	−9/17	: 74/17	x_{s1}
0 :	5/17	0	1	−1/17	0	3/17	−1/17	: 12/17	x_3
0 :	$\boxed{9/17}$	1	0	22/17	0	2/17	5/17	: 8/17	x_2 →
1 :	0	0	0	0	0	12/11	5/22	: 48/11	Z

(↑ under x_1)

Third Optimal Tableau

Z	x_1	x_2	x_3	x_4	x_{s1}	x_{s2}	x_{s3}	sol	BI
0 :	0	−28/9	0	−23/9	1	−7/9	−13/9	: 26/9	x_{s1}
0 :	0	−5/9	1	−7/9	0	1/9	−2/9	: 4/9	x_3
0 :	1	17/9	0	$\boxed{22/9}$	0	2/9	5/9	: 8/9	x_1 →
1 :	0	0	0	0	0	12/11	5/22	: 48/11	Z

(↑ under x_4)

The complete analysis of the LP problem shows that there are three basic (extreme point) primal optimal solutions and a unique dual solution corresponding to an optimal value of the objective function equal to 48/11. In each optimal tableau, the dual solution exhibits the degeneracy of two activities, a property that permits the discovery of the other optimal solutions. The pivot in each tableau indicates the path chosen to go from one optimal extreme point to the next one. The third optimal tableau leads to the first optimal tableau, repeating the cycle of optimal solutions. The primal basic optimal solutions are reported in table 15.1.

Table 15.1. Multiple Optimal Solutions

	Extreme Point #1	Extreme Point #2	Extreme Point #3
x_1	0	0	8/9
x_2	0	8/17	0
x_3	8/11	12/17	4/9
x_4	4/11	0	0

With this evidence available to the consultant what should she tell the vintner? At first sight, it appears that the right answer is to discontinue two of the four activities, that is, the vintner should produce only one type of wine and the corresponding bottles. The consultant, however, knows that the vintner likes to produce all three types of wine. She also remembers from her college studies that a convex combination of optimal solutions is itself an optimal solution. Hence, before writing the report with her recommendations, she attempts to give empirical meaning to the notion of a convex combination of optimal solutions. This means that choosing nonnegative weights β_1, β_2, and β_3 such that $\beta_1 + \beta_2 + \beta_3 = 1$, the following weighted sum of extreme point optimal solutions generates a nonextreme point optimal solution:

$$\begin{matrix} \text{nonextreme} \\ \text{point} \\ \text{optimal} \\ \text{solution} \end{matrix} \equiv \begin{bmatrix} x_1 \\ x_2 \\ x_3 \\ x_4 \end{bmatrix} = \beta_1 \begin{bmatrix} 0 \\ 0 \\ 8/11 \\ 4/11 \end{bmatrix} + \beta_2 \begin{bmatrix} 0 \\ 8/17 \\ 12/17 \\ 0 \end{bmatrix} + \beta_3 \begin{bmatrix} 8/9 \\ 0 \\ 4/9 \\ 0 \end{bmatrix}$$

Any combination of values of the weights β_1, β_2, $\beta_3 \geq 0$ that also add up to one generates an optimal solution. The consultant, therefore, decides to choose weights $\beta_i, i = 1, 2, 3$ such that they minimize the distance between the combination of activities produced by the vintner and an optimal combination of LP solutions as determined by her model. It seems plausible, in fact, that the vintner will more likely accept the consultant's conclusions if they do not require major changes from current practices.

The solution of this problem requires the estimation of a simple linear regression, a subject that is beyond the scope of this book. Nonetheless, we state such a problem in order to indicate a relatively simple approach to the problem of using the information from a linear programming problem to evaluate the economic behavior of economic agents. The linear regression required to find the weights β_i can be estimated using the least-squares method based upon the minimization of the sum of squared residuals between the observed level of wine and bottle activities and the corresponding combination of LP solutions:

$$\text{minimize} \sum_{j=1}^{4} e_j^2$$

$$\text{subject to } e_j + \beta_1 x_{1j} + \beta_2 x_{2j} + \beta_3 x_{3j} = \text{observed } x_j$$

$$\beta_1 + \beta_2 + \beta_2 = 1$$

$$\beta_i \geq 0, \quad i = 1, 2, 3; \quad j = 1, \ldots, 4$$

The "observed x_j" are the observed levels of the wine and bottle activities produced by the vintner, while the x_{1j}, x_{2j}, x_{3j}, are the jth components of the three optimal LP solutions from the consultant's model. It turns out that in this example the optimal value of the weights are $\beta_1 = 1/4$, $\beta_2 = 1/4$, and $\beta_3 = 1/2$. Hence the "best" (in the sense of minimizing the sum of squared residuals) weighted combination of the three optimal LP solutions is the following nonbasic solution:

$$
\begin{bmatrix} x_1 \\ x_2 \\ x_3 \\ x_4 \end{bmatrix} = \frac{1}{4} \begin{bmatrix} 0 \\ 0 \\ 8/11 \\ 4/11 \end{bmatrix} + \frac{1}{4} \begin{bmatrix} 0 \\ 8/17 \\ 12/17 \\ 0 \end{bmatrix} + \frac{1}{2} \begin{bmatrix} 8/9 \\ 0 \\ 4/9 \\ 0 \end{bmatrix} = \begin{bmatrix} 4/9 \\ 2/17 \\ 329/621 \\ 1/11 \end{bmatrix}
$$

The optimal combination of optimal LP solutions is equal to the observed levels of wine and bottle activities, and the best strategy for this vintner is to continue to do what he was already doing. Of course, this conclusion is rare and corresponds to an extreme case. In general, optimal and observed levels will differ and the advantage of the approach illustrated above is to minimize the difference between what an economic agent is doing and what ought to be done to maximize profits. Also, by adopting the above approach, it is possible to suggest an optimal strategy that is more diversified (has more activities at positive levels) than any primal basic optimal solution. This, to recall, is the meaning of a nonbasic solution. Finally, to verify that the nonbasic solution is nonetheless an optimal solution, we evaluate it in terms of the objective function coefficients to find that

$$\frac{53}{22}\frac{4}{9} + \frac{39}{22}\frac{2}{17} + 5\frac{329}{621} + 2\frac{1}{11} = \frac{48}{11} = Z_{\text{opt}}$$

corresponding to the optimal value of the objective function for each of the three basic optimal solutions.

15.3   Dual Multiple Optimal Solutions

Problems exhibiting dual multiple optimal solutions are not uncommon. From an economic viewpoint the existence of dual multiple optimal solutions means that an infinite number of shadow price systems for the limiting inputs are consistent with the goal of minimizing the value of resources. A numerical example will illustrate this proposition.

Consider the following linear programming problem:

Primal maximize $Z = 4x_1 + x_2 + 5x_3$

subject to
$$\begin{aligned}
3x_1 + 2x_2 - x_3 &\leq 2 \\
-x_1 + x_2 + 5x_3 &\leq 4 \\
x_1 + x_3 &\leq 2
\end{aligned}$$

$$x_i \geq 0, \qquad i = 1, \ldots, 3$$

There are two dual multiple basic optimal solutions for this example. The corresponding optimal tableaux are reported below:

First Optimal Tableau

Z	x_1	x_2	x_3	x_{s1}	x_{s2}	x_{s3}	sol P	BI
0 :	1	11/14	0	5/14	1/14	0 :	1	x_1
0 :	0	1/2	1	1/14	5/14	0 :	1	x_3
0 :	0	$\boxed{-8/7}$	0	$-6/14$	$-2/7$	1 :	0	x_{s3} →
1 :	0	55/14	0	25/14	19/14	0 :	9	Z
D sol	y_{s1}	y_{s2}	y_{s3}	y_1	y_2	y_3	R	

↑

Second Optimal Tableau

Z	x_1	x_2	x_3	x_{s1}	x_{s2}	x_{s3}	sol P	BI
0 :	1	0	0	1/16	−1/8	11/16 :	1	x_1
0 :	0	0	1	−13/112	13/56	7/16 :	1	x_3
0 :	0	1	0	1/8	1/4	$\boxed{-7/8}$:	0	x_2 →
1 :	0	0	0	5/16	3/8	55/16 :	9	Z
D sol	y_{s1}	y_{s2}	y_{s3}	y_1	y_2	y_3 ↑	R	

As gleaned from the tableaux, there is a unique primal solution corresponding to two optimal dual solutions. The algorithm to generate the second optimal solution is the dual simplex method. As asserted, the two dual solutions can be interpreted as two systems of shadow prices for the given limiting resources. The existence of multiple shadow prices for the same commodity is the dual counterpart to the existence of multiple optimal quantities of the same commodity for a given price, as discussed in the previous section. As before, therefore, we can generate an infinity of shadow price systems that are all optimal by taking convex combinations of the two basic dual optimal solutions. Thus, let $\beta_1 = 4/5$ and $\beta_2 = 1/5$ be nonnegative arbitrary weights that clearly add up to one. Then the following convex combination is also an optimal dual solution, although a nonbasic one:

$$\begin{bmatrix} y_1 \\ y_2 \\ y_3 \end{bmatrix} = \frac{4}{5} \begin{bmatrix} 25/14 \\ 19/14 \\ 0 \end{bmatrix} + \frac{1}{5} \begin{bmatrix} 5/16 \\ 3/8 \\ 55/16 \end{bmatrix} = \begin{bmatrix} 167/112 \\ 65/56 \\ 11/16 \end{bmatrix}$$

The optimality of the new dual solution is verified by noticing that it is nonnegative, it satisfies the dual constraints, and it reproduces the optimal value of the dual objective function:

$$2\frac{167}{112} + 4\frac{65}{56} + 2\frac{11}{16} = 9 = \min R$$

This example indicates that in order to have dual multiple optimal solutions it is necessary that the primal solution be degenerate. In the first optimal tableau, in fact, the basic variable x_{s3} is equal to zero, as is x_2 in the second optimal tableau. The search for optimal weights to be utilized in the convex combination of basic optimal dual solutions can be conducted by means of a least-squares procedure as discussed in the previous section.

15.4 Problems with Multiple Primal and Dual Optimal Solutions

In general, multiple optimal solutions loom in the background of algorithms and only a determined effort will bring them to the forefront. Unfortunately, all available commercial computer programs for solving mathematical programming problems have not addressed this issue. They have not even attempted to do so. In spite of this unjustified omission, the thoughtful researcher must be aware of the opportunity to enrich her final report with a series of efficient alternatives incorporated in the information organized in a LP problem. To further illustrate the possibilities, we discuss a plausible and simple linear programming problem that exhibits both primal and dual multiple optimal solutions. Consider the following LP problem:

$$\textbf{Primal} \ \text{maximize} \ Z \ = \ \frac{53}{22}x_1 \ + \ \frac{5}{3}x_2 \ + \ 5x_3 \ + \ 2x_4$$

$$\text{subject to} \quad \frac{121}{55}x_1 \ + \ 2x_2 \ + \ x_3 \ + \ 4x_4 \ \le \ \frac{24}{11}$$
$$2x_1 \ + \ x_2 \ + \ 5x_3 \ + \ x_4 \ \le \ 4$$
$$x_1 \ + \ 3x_2 \ - \ 2x_3 \ + \ 4x_4 \ \le \ 0$$

$$\text{all} \ x_j \ge 0$$

This problem has two optimal basic primal solutions and two optimal basic dual solutions. The following three tableaux contain such solutions and indicate the path connecting them by means of the boxed pivots.

First Optimal Tableau
Dual simplex algorithm

	Z	x_1	x_2	x_3	x_4	x_{s1}	x_{s2}	x_{s3}	sol P	BI
0 :	0	$\boxed{-25/22}$	0	0	1	$-6/11$	$-19/22$	: 0*	$x_{s1} \rightarrow$
0 :	7/22	1/22	1	0	0	2/11	$-1/22$	: 8/11	x_3
0 :	9/22	17/22	0	1	0	1/11	5/22	: 4/11	x_4
1 :	0*	7/66	0	0	0	12/11	5/22	: 48/11	Z
D sol	y_{s1}	y_{s2}	y_{s3}	y_{s4}	y_1	y_2	y_3	R	
		↑							

Second Optimal Tableau
Primal simplex algorithm

Z	x_1	x_2	x_3	x_4	x_{s1}	x_{s2}	x_{s3}	sol P	BI
0 :	0	1	0	0	$-22/25$	$12/25$	$19/25$:	0^*	x_2
0 :	$7/22$	0	1	0	$1/25$	$4/25$	$-2/25$:	$8/11$	x_3
0 :	$\boxed{9/22}$	0	0	1	$17/25$	$-7/25$	$-9/25$:	$4/11$	$x_4 \rightarrow$
1 :	0^*	0	0	0	$7/75$	$78/75$	$11/75$:	$48/11$	Z

D sol $\quad y_{s1} \quad y_{s2} \quad y_{s3} \quad y_{s4} \quad y_1 \quad y_2 \quad y_3 \qquad R$
$\qquad\quad \uparrow$

Third Optimal Tableau
Primal and dual simplex algorithms

Z	x_1	x_2	x_3	x_4	x_{s1}	x_{s2}	x_{s3}	sol P	BI
0 :	0	1	0	0	$\boxed{-22/25}$	$12/25$	$19/25$:	0^*	$x_2 \rightarrow$
0 :	0	0	1	$-7/9$	$-22/45$	$9/25$	$1/5$:	$4/9$	x_3
0 :	1	0	0	$\boxed{22/9}$	$374/225$	$-154/225$	$-22/25$:	$8/9$	$x_1 \rightarrow$
1 :	0	0	0	0^*	$7/75$	$78/75$	$11/75$:	$48/11$	Z

D sol $y_{s1} \quad y_{s2} \quad y_{s3} \quad y_{s4} \qquad y_1 \qquad y_2 \qquad y_3 \qquad R$
$\qquad\qquad\qquad\quad \uparrow \qquad\quad \uparrow$

The locations of the degenerate solutions are indicated with an asterisk (*). By following the pivots, it is possible to trace out the path from one extreme point to the next. The third optimal tableau exhibits two pivots that allow returning to either the first or the second optimal tableau. Notice that either dual solution can be associated with both primal solutions and vice versa.

The discussion has centered around the importance of multiple optimal solutions and how to compute them. The examples illustrate the feasibility of exploring a linear programming problem for the possibility of identifying either some or all the optimal solutions. Surprisingly, the commercially available computer routines for solving linear programs do not incorporate the option of computing multiple optimal solutions when they exist.

15.5 Modeling to Generate Alternatives

The previous sections have discussed the meaning and uses of multiple optimal solutions. In this section we go one step further and present arguments in favor of using linear programming as a framework to generate potentially interesting solutions that will have to be ultimately evaluated by the decision maker. It is not unlikely, in fact, that the optimal solution of a linear programming problem, particularly when unique, would be too specialized to be of any interest for the entrepreneur who has commissioned the study. When this event occurs, disappointment ensues. Yet the LP consultant is well aware that the amount of information that has gone into the formulation of the system of LP constraints is disproportionate to the result as judged by the decision maker. The problem, then, reduces to finding ways for the inputted information to transpire in the form of strategies well worth consideration. If the optimal solution does not attract the attention of the decision maker who judges it irrelevant, it is worthwhile to trade off some optimality in order to achieve some relevance. As usual, an example will help in clarifying the point.

Consider Stigler's diet problem once again. In section 12.1 we saw that the annual minimum cost (at 1939 prices) for an adequate diet could have been as low as $39.66. The minimum cost diet, however, imposed stringent boundaries to its palatability. Only 5 foods were chosen among the available 15 to be eaten in an entire year, admittedly a not too cheerful prospect. It is clear, therefore, that a single linear programming solution may not be the desired information on which to base one's actions. One senses that the information about the content and prices of food requirements contained in table 1.2 was not processed in a desirable manner, and most of it remained hidden in the LP model. Thirty-nine dollars constituted a rather cheap diet even for the year 1939. Let us see, therefore, what is possible to do in terms of increasing the variety of foods by gradually increasing the allowable spending limit. Table 15.2 presents the next minimum cost of daily diets corresponding to various percent increases over the true minimum based upon the information of table 1.2 involving 15 foods. With a 2 percent increase, evaporated milk and cheese enter the solution, while liver is eliminated. At 5 percent, the composition of the foods remains the same with some variation in the levels of consumption. The major shift here is from navy beans to wheat flour. At 10 percent, oleomargarine enters the diet at the expense of navy beans. At the 20, 30, and 50 percent increases, spinach is reduced to an insignificant level or eliminated and liver reenters the solution. At an increase of 100 percent over the minimum cost, potatoes appear, while flour is reduced to about half a pound per day. Over the range of 100 percent increase of the diet cost, which now has reached the

Table 15.2. Diets for Various Percent Increases over the Minimum Cost Based on the 15 Foods of Table 1.2 (lb/day)

Food	Least Cost Diet	Percent Increase						
		2%	5%	10%	20%	30%	50%	100%
Wheat flour	0.82	1.18	1.54	1.54	1.29	1.11	0.82	0.43
Evaporated milk	–	0.16	0.21	0.37	0.64	0.64	0.76	1.18
Oleomargarine	–	–	–	0.05	0.16	0.25	0.36	0.50
Cheese	–	0.03	0.09	0.07	–	–	–	–
Liver	0.01	–	–	–	0.01	0.01	0.01	0.01
Cabbage	0.30	0.31	0.31	0.32	0.33	0.34	0.33	0.07
Onions	–	–	–	–	–	–	–	–
Potatoes	–	–	–	–	–	–	–	0.80
Spinach	0.6	0.6	0.05	0.05	0.03	0.01	–	–
Sweet potatoes	–	–	–	–	–	–	–	–
Dried peaches	–	–	–	–	–	–	–	–
Dried prunes	–	–	–	–	–	–	–	–
Dried beans (Lima)	–	–	–	–	–	–	–	–
Dried beans (Navy)	1.03	0.58	0.14	–	–	–	–	–
Daily cost (cts)	10.87	11.08	11.41	11.95	13.04	14.13	16.30	21.73
Annual cost ($)	39.66	40.46	41.65	43.63	47.59	51.56	59.49	79.32

level of 21.73 cents per day or \$79.32 per year, the LP model indicates that it is possible to count on 9 different foods (out of 15) to produce a diet of adequate nutrition.

The precise linear programming specification leading to the discovery of the alternative solutions is indicated in symbolic notation as follows:

$$\text{minimize} \quad x_{s0}$$

$$\text{subject to} \quad -x_{s0} + \sum_{j=1}^{n} c_j x_j = (1 + pc)MTC$$

$$\sum_{j=1}^{n} a_{ij} x_j \geq b_i$$

$$x_j \geq 0, \quad j = 1, \ldots, n$$

where MTC is the minimum total cost of the optimal diet, pc is the allowed percentage increase in the total cost set to discover alternative near optimal

Table 15.3. Diets for Various Percent Increases over the Minimum Cost Based on the 77 Foods of Table 1.2 and Table 12.3 (lb/day)

Food	Least Cost Diet	Percent Increase						
		2%	5%	10%	20%	30%	50%	100%
Wheat flour	0.82	1.18	1.54	1.54	1.29	1.11	0.82	0.43
Wheat cereals	–	–	–	–	–	0.08	–	–
Corn meal	–	–	–	–	–	–	–	0.48
Milk	–	–	–	–	–	–	0.66	0.60
Evaporated milk	–	0.16	0.21	0.66	0.68	0.37	–	–
Crisco	–	–	–	–	–	–	–	0.38
Liver (beef)	0.01	–	–	0.01	0.03	0.04	0.03	0.07
Potatoes	–	–	–	–	–	–	0.06	0.06
Spinach	0.06	0.06	0.05	0.05	0.05	0.03	0.03	–
Dried beans (Lima)	–	–	–	–	0.01	0.16	0.05	–
Dried beans (Navy)	1.04	0.58	0.13	–	–	–	–	–
Sugar	–	–	–	–	0.04	0.07	0.06	–
Cabbage	0.30	0.31	0.31	0.31	0.30	0.30	–	–
Lard	–	–	–	0.15	0.11	–	–	–
Cheese	–	0.03	0.09	–	–	–	–	–
Daily cost (cts)	10.87	11.08	11.41	11.95	13.04	14.13	16.30	21.73
Annual cost ($)	39.66	40.46	41.65	43.63	47.59	51.56	59.49	79.32

solutions, b_i is the minimum requirement for the ith nutrient, c_j is the price of the jth food, a_{ij} is the content of the ith nutrient in the jth food, and x_{s0} is the slack variable of the cost constraint that should be minimized. The objective function of the original diet problem, in other words, has become a new constraint. The objective is now to find an adequate and economically efficient diet that is only slightly more costly than the true minimum diet by an amount specified in the new constraint.

When the same parametric variation of the objective function is carried out using all the 77 foods of table 1.2 and table 12.3, further interesting results are obtained. The combinations of foods appearing in the various optimal solutions are given in table 15.3. Now a larger variety of food items appears convenient and justifies the wide variety of foods with which Stigler began the analysis. Two solutions, corresponding to an increase of the diet cost of 20 and 30 percent, include 8 categories of food. Overall, 15 different foods appear in the table, a significant improvement in the palatability of an adequate and reasonably thrifty diet.

In summary, the procedure of relaxing the optimal level of the objective function by some percentage points is a sensible way to explore alternative solutions that, although not strictly optimal, may be appealing on other grounds not reflected into the linear programming model.

This increased diversity of the primal solutions near the optimal point has a price: we can no longer rely on the availability of the dual variables as economic indicators of the relative tightness of the nutrient requirement constraints. Now that the primal solution is no longer optimal (it is only near optimal), the dual solution is infeasible. This is a consequence of the discussion carried out in chapter 9 about the Lagrangean function and the complementary slackness conditions. To see how this infeasibility comes about, recall that feasible primal and dual solutions are optimal when the corresponding objective functions are equal. Suppose x^* and y^* are optimal solutions: then,

$$\sum_{j=1}^{n} c_j x_j^* = \sum_{i=1}^{m} b_i y_i^*$$

At this point, however, we elect to settle for a diet whose costs are a few percentage points higher than the optimal solution. Let us call this near optimal diet $\overline{x}$. The relationship between the primal and dual objective functions is now

$$\sum_{j=1}^{n} c_j \overline{x}_j > \sum_{i=1}^{m} b_i y_i^*$$

and the strict inequality prevents the dual objective function from achieving the same value of the primal counterpart. This is an indication that feasible levels of the dual variables do not exist to match the levels of the corresponding primal variables and, therefore, the dual problem is infeasible. This is precisely the meaning of *near optimality:* when the primal solution is not optimal, the corresponding dual solution is infeasible. From the duality relations studied in previous chapters we can reverse the conclusion: when the dual solution is infeasible, the corresponding primal solution is not optimal. This is the price to pay for accepting near optimality.

We must recall, finally, that one simple way to obtain near optimal solutions is to scrutinize the opportunity cost of those activities not in the optimal basis. If the corresponding $(z_j - c_j)$ is judged to be sufficiently small (relative to the c_j), an increase (in case of a maximization) of c_j by the amount indicated by the $(z_j - c_j)$ will characterize the jth activity by the break-even property and make it suitable to enter the basis.

Chapter 16

Solid Waste Management

16.1   A Growing Problem

Solid waste in urban settlements has become a principal problem and headache of local governments around the world. Until a couple of decades ago, the solution consisted of collecting and transporting the waste from urban areas to landfills where it was simply buried. Increasing environmental consciousness, coupled with a limited supply of suitable landfills, has raised the price of waste disposal to a level that makes it convenient to look at solid waste not simply as worthless refuse but as a source of raw materials and energy. The growth of modern economies has occurred in association with consumerism, which, in turn, has spurred a mentality of discarding goods containing considerable amounts of recoverable resources, from paper to glass, to metals, and finally, to energy (potential fuel). A new economic sector devoted to recycling was thus born. In spite of private entrepreneurship and ingenuity, the management of solid waste is a daily concern of local governments for a relentlessly growing problem that risks suffocating communities, especially when the disposal network breaks down for even a few days.

The management of solid waste is a real-life problem. From our perspective, therefore, it is important to examine what linear programming can offer in the form of an analytical framework capable of organizing the many variables and relations involved.

16.2   A LP Model of Waste Management

We present a simplified but instructive example of solid waste management based on a study by Julia Friedman. The scene is Lane County in western Oregon around 1974. The twin towns of Eugene-Springfield and the territory within a 20-mile radius produce 700 tons per day of solid waste or 255,500 tons per year. The disposal of this waste is a public as well as a private concern. For example, charitable, commercial, and industrial enterprises sort a portion of the total waste for materials (paper, glass, ferrous, and other items) that have a resale value on secondary markets. No aluminum recovery is contemplated in this model because the major source of aluminum (in the form of tin cans) has been removed from solid waste by the famous Oregon beverage container law that imposes a 5-cent deposit per can. The private handling of waste is estimated at a maximum of 10 percent of the total. The remaining portion is assumed to be handled directly by a public agency at one central plant where a number of recovery activities are possible. The waste tonnage enters the facility either for transfer to the landfill or for processing. It can be hand sorted for recovering paper and cardboard, shredded and sorted by air, further subjected to magnetic separation for recovering ferrous materials and, finally, treated for energy recovery. Paper and cardboard, ferrous materials, and fuel so produced are sold on secondary markets.

It is clear that some of the activities listed above imply a net cost, while others might be associated with a net revenue. We assume that the objective is to minimize the total cost of disposing of the solid waste in the Eugene-Springfield area, whether by private or by public organizations.

Following Friedman, the detailed list of activities entering the linear programming model is specified as follows:

x_0 ≡ total annual solid waste for the Eugene-Springfield area

x_1 ≡ tons of solid waste annually sorted at private sites (home, commercial, charitable) for resale to secondary markets

x_2 ≡ tons of solid waste annually hauled from the urban area to the central facility either for transfer to the landfill or for processing

x_3 ≡ tons of solid waste annually hand-sorted at the center to recover materials for resale on secondary markets

x_4 ≡ tons of solid waste annually shredded and air classified at the processing plant

x_5 ≡ tons of solid waste annually removed by magnetic separation for sale in secondary markets

x_6 ≡ tons of solid waste annually hauled from the processing center to the landfill

x_7 ≡ tons of solid waste annually recovered for fuel production

The technological relations describing the various options for the management of solid waste are divided among processing, material balance, and bounding constraints.

Processing constraints. Four relations are specified:

1. Separation of recyclable material by private enterprises:

$$x_1 \leq p_s x_0$$

where p_s is the proportion of solid waste that can be removed prior to delivery to the central plant. For this study $p_s = 0.1$.

2. Hand separation of recyclable materials:

$$x_1 + x_3 \leq p_h x_0$$

where p_h is the proportion of x_0 that can be removed either by private enterprise or by hand sorting at the central plant. The coefficient $p_h = 0.12$.

3. Magnetic recovery of ferrous materials:

$$p_{f1}x_1 + p_{f3}x_3 + p_{f5}x_5 \leq p_f x_0$$

where p_f is the fraction of recoverable ferrous metal in the total annual solid waste and p_{fj} are the fractions of the same metals recoverable from the jth activity. In this study, $p_f = 0.06$, $p_{f1} = 0.1$, $p_{f3} = 0.5$, and $p_{f5} = 1$.

4. Recovery of fuel material:

$$p_{e1}x_1 + p_{e3}x_3 + p_{e7}x_7 \leq p_e x_0$$

where p_e is the proportion of potential fuel in the total annual solid waste load and p_{ej} are fractions of potential fuel in the jth activity. The various fractions are $p_e = 0.65$, $p_{e1} = 0.7$, $p_{e3} = 0.25$, and $p_{e7} = 1$.

Material balance constraints. There are three such constraints:

5. The materials separated by private enterprises plus those hauled to the central plant must equal the total annual solid waste load:

$$x_1 + x_2 = x_0$$

6. The total weight of material entering the processing plant must equal the amount hand separated plus the amount shredded and air classified:

$$x_2 = x_3 + x_4$$

7. The total weight of material hauled to the processing plant must equal the sum of materials hand separated, magnetically separated, landfilled, and sorted for fuel recovery:

$$x_2 = x_3 + x_5 + x_6 + x_7$$

Bounding constraints. There are two such constraints:

8. The maximum fuel value extracted cannot exceed the fuel content of the processed materials:

$$x_7 \leq .7x_4$$

9. The amount of ferrous scrap magnetically extracted cannot exceed the scrap content of processed materials:

$$x_5 \leq .04x_4$$

The objective function of this linear programming problem intends to minimize the total net cost (*TNC*) of solid waste management, where net cost is the difference between the cost of the activity and the value of the associated recoverable materials. For the 20-mile radius of the Eugene-Springfield area, the coefficients of the total net cost function (measured in dollars per ton) were based upon a local survey of the various processing activities. The primal version of the solid waste management problem can thus be stated as follows:

Primal

$$\min TNC = -.1x_1 + 3.33x_2 - 3.0x_3 + 1.81x_4 - 9.2x_5 + 4.07x_6 - 1.59x_7$$

subject to

1.	x_1					$-.10x_0$	$\leq$	0
2.	x_1	$+ x_3$				$-.12x_0$	$\leq$	0
3.	$.1x_1$	$+ .5x_3$	$+ x_5$			$-.06x_0$	$\leq$	0
4.	$.7x_1$	$+ .25x_3$		$+ x_7$		$-.65x_0$	$\leq$	0
5.	$x_1 + x_2$					$- x_0$	$=$	0
6.	$x_2 -$	$x_3 -$	x_4				$=$	0
7.	$x_2 -$	x_3	$- x_5 - x_6 - x_7$				$=$	0
8.		$- .7x_4$	$+ x_7$				$\leq$	0
9.		$- .04x_4 + x_5$					$\leq$	0
10.						x_0	$=$	$255{,}500$

all variables are nonnegative

Positive coefficients of the objective function represent unit costs, while negative coefficients are unit net revenues. This formulation of the primal problem, with the variable x_0 representing total solid waste appearing explicitly among the active variables, is convenient for the parametric analysis of the total solid waste load. The supply of solid waste, x_0, in fact, enters into the determination of the first four constraints and cannot be varied independently of them. It turns out, as we shall see shortly, that the amount of total solid waste in this particular problem can range from zero to infinity without requiring a basis change. For this reason, this waste management problem can have the following alternative primal specification:

Primal

min $TNC = -.1x_1 + 3.33x_2 - 3.0x_3 + 1.81x_4 - 9.2x_5 + 4.07x_6 - 1.59x_7$

subject to

1.	x_1							$\leq$	25,550
2.	x_1	$+$	x_3					$\leq$	30,660
3.	$.1x_1$	$+$	$.5x_3$	$+ x_5$				$\leq$	15,330
4.	$.7x_1$	$+$	$.25x_3$				$+ x_7$	$\leq$	166,075
5.	$x_1 + x_2$							$=$	255,500
6.		$x_2 -$	$x_3 -$	x_4				$=$	0
7.		$x_2 -$	x_3		$- x_5 - x_6 - x_7$			$=$	0
8.				$- .7x_4$			$+ x_7$	$\leq$	0
9.				$- .04x_4 + x_5$				$\leq$	0

all variables are nonnegative

The right-hand-side quantities of the first four constraints are computed using the specified proportions and an amount of solid waste of 255,500 tons. It is now clearer that a parametric analysis of any one constraint in this second specification would be improper due to the structural linkage between the total solid waste and the various fractions of it represented by the first four constraints.

16.3   The Dual Problem

The dual specification of this solid waste management problem presents some interesting features. Notice, in fact, that the primal specification is stated as a minimization subject to *less-than-or-equal constraints*. Throughout this book, however, we have emphasized that a minimization structure is better associated with *greater-than-or-equal constraints*. This solid waste problem is an illustration of the importance of following that recommendation. With the presence of equality constraints among the primal restrictions, it is not immediately obvious what the proper sign of the dual coefficients should be. An incorrect sign of such coefficients can easily cause either infeasibility or unboundedness of the dual solution. The safe approach to the specification of the dual problem is, therefore, to begin with a primal specification stated in the form of a minimization subject to greater-than-or-equal constraints. At the cost of appearing redundant, we will state also the equality restrictions in that form. The counterproof of the importance of this operation is left to the reader who is challenged to write down the dual problem (and solving it) following directly the primal specifications of the previous section. For our part, we begin the formulation of the dual problem with a further specification of the primal according to the standard form:

Primal

min $TNC = -.1x_1 + 3.33x_2 - 3.0x_3 + 1.81x_4 - 9.2x_5 + 4.07x_6 - 1.59x_7$

subject to

										dual variables
1.	$- \quad x_1$						$\geq$	$-25,550$		y_1
2.	$- \quad x_1$	$- \quad x_3$					$\geq$	$-30,660$		y_2
3.	$- .1x_1$	$- .5x_3$		$- x_5$			$\geq$	$-15,330$		y_3
4.	$- .7x_1$	$- .25x_3$				$- x_7$	$\geq$	$-166,075$		y_4
5a.	$- \quad x_1 - x_2$						$\geq$	$-255,500$		y_5^a
5b.	$x_1 + x_2$						$\geq$	$255,500$		y_5^b
6a.	$- x_2 +$	$x_3 +$	x_4				$\geq$	0		y_6^a
6b.	$x_2 -$	$x_3 -$	x_4				$\geq$	0		y_6^b
7a.	$- x_2 +$	x_3		$+ x_5 + x_6 + x_7$			$\geq$	0		y_7^a
7b.	$x_2 -$	x_3		$- x_5 - x_6 - x_7$			$\geq$	0		y_7^b
8.			$.7x_4$			$- x_7$	$\geq$	0		y_8
9.			$.04x_4 - x_5$				$\geq$	0		y_9

all variables are nonnegative

The equality restrictions have been replaced by two inequality constraints according to the discussion of section 2.2. The dual variables associated with this formulation of the primal constraints are all nonnegative. The dual problem can now be stated correctly following the conventions discussed in chapter 2. The objective is to maximize the total net value (*TNV*) of solid waste in its various classifications:

Dual

Max $TNV = -25,550y_1 - 30,660y_2 - 15,330y_3 - 166,075y_4 - 255,500(y_5^a - y_5^b)$

subject to

1. $- y_1 - y_2 - .1y_3 - .7y_4 - (y_5^a - y_5^b)$ $\leq$ $-.1$
2. $- (y_5^a - y_5^b) - (y_6^a - y_6^b) - (y_7^a - y_7^b)$ $\leq$ 3.33
3. $- y_2 - .5y_3 - .25y_4 + (y_6^a - y_6^b) + (y_7^a - y_7^b)$ $\leq$ -3.0
4. $(y_6^a - y_6^b) + .7y_8 + .04y_9 \leq$ 1.81
5. $- y_3 + (y_7^a - y_7^b) - y_9 \leq$ -9.2
6. $(y_7^a - y_7^b) \leq$ 4.07
7. $- y_4 + (y_7^a - y_7^b) - y_8 \leq$ -1.59

all dual variables are nonnegative

The final form of the dual problem is obtained by defining the following unrestricted dual variables: $y_5 = (y_5^a - y_5^b)$, $y_6 = (y_6^a - y_6^b)$, and $y_7 =$

$(y_7^a - y_7^b)$. Then,

Dual

max $TNV = -25{,}550y_1 - 30{,}660y_2 - 15{,}330y_3 - 166{,}075y_4 - 255{,}500(y_5^a - y_5^b)$

subject to

1. $-y_1 - y_2 - .1y_3 - .7y_4 - y_5$			$\leq$	$-.1$
2. $-y_5 - y_6 - y_7$			$\leq$	3.33
3. $-y_2 - .5y_3 - .25y_4 + y_6 + y_7$			$\leq$	-3.0
4. $y_6 + .7y_8 + .04y_9$			$\leq$	1.81
5. $-y_3 + y_7 - y_9$			$\leq$	-9.2
6. y_7			$\leq$	4.07
7. $-y_4 + y_7 - y_8$			$\leq$	-1.59

y_5, y_6, and y_7 are free, all other dual variables are nonnegative

When scrupulously followed, this seemingly lengthy process will actually save time and difficulties in the solution and interpretation of the dual problem.

The economic interpretation of the dual constraints follows the general scheme of marginal cost being greater than or equal to marginal revenue. According to this framework, therefore, the first dual constraint should be read as

$$y_1 + y_2 + .1y_3 + .7y_4 + y_5 \geq .1$$

where .1 is the marginal revenue of the private processing activity. Similarly, the last dual constraint should be read after a rearrangement such as

$$y_4 + y_8 \geq 1.59 + y_7$$

which states that the total marginal cost of the fuel recovery activity is composed of two parts: the marginal cost due to the technological relation between solid waste in general and fuel content, y_4, plus the marginal cost of the bound on extractable fuel from shredded and air classified waste, y_8. The imputed marginal revenue of the same activity is the sum of the net price for fuel obtained on the market, \$1.59, plus the marginal value of the solid waste hauled to the central plant and processed for fuel content, y_7. Analogous economic interpretation can be formulated for the remaining dual constraints.

16.4   Solution of the Waste Management Problem

The primal and dual solutions of the solid waste management problem are presented in table 16.1 together with the corresponding sensitivity or range

Table 16.1. Solution and Range Analysis of the Waste Management
Problem (units of hundred thousand tons)

Primal Solution	Level	Range Analysis of Objective Coefficients		
		Lower range	Current	Upper range
x_1 = private source	0.22484	−2.217	−0.10	−3.091
x_2 = hauling	2.33016	0.139	3.33	5.447
x_3 = hand sorting	0.08176	−6.191	−3.00	−0.883
x_4 = shredding	2.24840	−2.689	1.81	infinity
x_5 = magnetic sorting	0.08994	−121.689	−9.20	−1.223
x_6 = landfill	0.67554	−0.634	4.07	67.878
x_7 = fuel	1.48292	7.453	−1.59	3.114
x_0 = total solid waste	2.55500	0.000	0.00	infinity

Dual Solution ($/ton)	Value	Range Analysis of RHS Coefficients		
		Lower range	Current	Upper range
y_1 = waste for private	0.0000	−0.03066	0.00	infinity
y_2 = hand-sorted material	4.4996	0.19551	0.00	0.02666
y_3 = ferrous materials	5.2926	−0.01226	0.00	0.08085
y_4 = solid waste for fuel	5.6600	−1.48292	0.00	0.09096
y_5 = material balance	−8.8909	−0.12209	0.00	0.30660
y_6 = shredding balance	1.4909	−0.30660	0.00	0.12209
y_7 = hauling balance	4.0700	−infinity	0.00	0.67554
y_8 = fuel upper bound	0.0000	−0.09096	0.00	infinity
y_9 = ferrous upper bound	7.9773	−0.08085	0.00	0.01226
y_0 = total solid waste	−4.3544	0.00000	2.555	infinity

Minimum Total Net Cost = $1,112,547.21

analysis. The primal variables are measured in units of hundred thousand
tons. As explained in chapter 7, this scaling is mandatory whenever a computer program is used for the numerical solution of a problem. Improper
scaling may cause large rounding-off errors and often prevents the achievement of a solution. The first primal formulation of section 16.2 was used
in the computations.

The solid waste management program in the Eugene-Springfield area
in 1974 could have had an annual net cost of $1,112,547 corresponding to
$4.3544 per ton. The division of the waste load between private and central
facilities shows 22,484 tons handled by private enterprises and 233,016 tons
hauled and processed by the central plant. The total cost of simply transferring 233,016 tons of solid waste to the central location and directly to the
landfill, without any processing, amounts to $1,724,318.4 = (3.33 + 4.07)
× 233,016. Through the solid waste management program, therefore, the

community could have saved \$611,771.40 annually. All the processing activities (hand sorting, shredding, magnetic separation, and fuel production) enter the solution at positive levels.

The dual solution indicates the imputed value per ton of the various fractions of solid waste hauled, processed, and landfilled. A ton of solid waste removed either by hand sorting at the central plant or by separation at private locations has an imputed value of \$4.50 per ton (y_2). The ferrous materials have a shadow price of \$5.29 per ton (y_3). The solid waste convertible to fuel has an imputed value of \$5.66 per ton. The necessity of disposing of all the solid waste accumulated in the Eugene-Springfield area (constraint 5) imposes a gross imputed cost of \$8.89 per ton (y_5). It becomes a net cost of \$4.35 per ton (y_0) when the revenue from the processing activities is taken into account. This conclusion is demonstrated by writing down the dual constraint of the primal activity corresponding to total solid waste, x_0, that is

$$y_0 = .1y_1 + .12y_2 + .06y_3 + .65y_4 + y_5$$
$$= 0.0 + 0.54 + 0.32 + 3.68 - 8.89 = -4.35$$

This imputed price, of course, is the same value obtained by dividing the total cost of the solid waste management program by the total waste load of 255,500 tons, that is, \$1,112,547/255,500 tons = \$4.35 per ton. The shredding balance has a net value of \$1.49 per ton (y_6), while the bound for the recovery of ferrous material shows a value of \$7.98 per ton (y_9).

The sensitivity (or range) analysis of the objective function coefficients as well as of the constraints, reported in table 16.1, indicates that the optimal solution is quite stable both on the primal as well as the dual side. Considering first the primal constraints, we discover that the supply of solid waste indicated by the constraint on x_0, can change from zero to infinity without requiring a change in the basis. This is the crucial information that allows the reduction of this problem to the second primal specification given above. The range analysis of the objective function coefficients reveals that the optimal solution is also stable with respect to the prices of the various activities. The net revenue of private enterprises can vary from −\$2.22 to \$3.09 per ton, and they would handle the same quantity of solid waste. Commercial hauling and transferring can vary from \$0.14 to \$5.45 per ton. The price (net revenue) for hand-separated materials can vary from −\$6.19 to −\$0.88 per ton while maintaining the same level of operation (8,176 tons). The price of shredding and air classification of solid waste can vary from a net revenue of −\$2.69 per ton to an infinite unit cost without requiring any changes in the optimal solution. The same conclusion can be drawn with respect to the prices of magnetic separation, landfilling, and fuel production. The conclusion is that the solution recommended by the linear programming model for the management of solid waste in Lane County, Oregon, is robust with respect to price variations of any activity involved. This fact diminishes any apprehension about the precision of price estimates from the local survey that could have risen initially.

16.5   Parametric Analysis of Solid Waste Management

The derivation of input demand and output supply functions of the various commodities involved in the solid waste management model can properly be derived by a full parametric analysis as discussed in chapters 13 and 14. The range analysis of the primal constraint corresponding to the total solid waste has already indicated that the waste load can vary from zero to infinity without requiring a basis change. Parametric analysis of the primal constraints is, therefore, precluded for this problem. We will present, then, the parametric analysis for each coefficient of the objective function, although we already know that the solution is very stable. In order to guarantee a sufficient exploration of the range of values for the objective function's coefficients, we have chosen an interval from -100 to $+100$, even though it may be difficult to assign economic meaning to each segment.

Realistically, parametric programming, even for a problem of such dimensions as the one discussed in this chapter, requires access to an efficient and complete computer program. For the analysis of this problem the program LINDO was used, which allows only the parametric evaluation of the right-hand-side coefficients. To perform the parametric analysis of the objective function's coefficients, therefore, the solid waste problem was inputted in its dual specification. The result is that the breaking points of the objective function's parameters are now associated with dual variables entering and exiting the dual basic solution rather than primal variables entering and exiting the primal basic solution. The results are reported in table 16.2. Boxed coefficients are the initial values used in the problem.

From table 16.2 we observe that private enterprises will operate at a level of $x_1 = 0.22484$ for an interval of net revenue that ranges between $-\$2.217$ and $\$3.091$ per ton. That is, even if those enterprises were to lose $\$3.091$ per ton, the necessity of disposing of all the solid waste in Lane County would make it optimal (in a least-cost sense) to operate the private activities at the same level as when they can reap a profit of $\$2.217$ per ton. The $\$3.091$ per ton can be thought of as the upper limit of a subsidy paid by the community to private enterprises for their survival in the business of disposing the annual 22,484 tons of solid waste. When the cost of this activity ranges between $\$3.091$ and $\$3.921$ per ton, the annual volume of solid waste diminishes to 2,271 tons.

Table 16.2. Parametric Analysis of the Objective Function Coefficients

Activity	Entering Variable	Exiting Variable	Objective Coefficient Value ($/ton)	New Value of Variable (100,000 tons)	Objective Function Value (100,000 $)
x_1, private			−100.000	0.25550	−14.3341
sorting	y_1	y_3	− 2.217	0.22484	10.6495
			$\boxed{-0.100}$	0.22484	11.1255
	y_8	y_9	3.091	0.22484	11.8429
	y_{s2}	y_2	3.921	0.02271	11.8618
			100.000	0.00000	11.8618
x_2, hauling			−100.000	2.55500	−252.1460
	y_5^a	y_5^b	4.589	2.55500	−8.3709
	y_{s2}	y_2	−0.691	2.53229	1.5881
	y_8	y_9	0.139	2.33016	3.6899
			$\boxed{3.330}$	2.33016	11.1255
	y_1	y_3	5.447	2.33016	16.0584
			100.000	2.29950	233.4830
x_3, hand sorting			−100.000	0.30660	−17.8394
	y_{s2}	y_2	−24.822	0.30660	5.21016
	y_{s6}	y_4	8.738	0.28389	10.14150
	y_6^b	y_6^a	−7.354	0.28389	10.53430
	y_8	y_9	−6.191	0.08176	10.86460
			$\boxed{3.000}$	0.08176	11.12550
	y_1	y_3	−0.883	0.08176	11.29860
	y_{s4}	y_2	3.934	0.05110	11.54470
			100.000	0.00000	11.54470
x_4, shredding			−100.000	2.55500	−247.33900
	y_5^a	y_6^b	−6.869	2.55500	−9.39984
	y_{s4}	y_3	−5.121	2.44391	−4.94137
	y_{s2}	y_2	−2.689	2.24840	1.00846
	y_6^b	y_6^a	0.319	2.24840	7.77333
			$\boxed{1.810}$	2.24840	11.12550
			100.00	2.24840	231.89600
x_5, magnetic			−100.000	0.08994	2.95928
sorting	y_6^a	y_6^b	−46.477	0.08994	7.77330
			$\boxed{-9.200}$	0.08994	11.12550
	y_8	y_9	−1.223	0.08994	11.84290
	y_6^b	y_6^a	1.686	0.00908	11.86940
	y_3	y_{s6}	4.070	0.00908	11.89100
			100.000	0.00000	11.89100

(continued)

Table 16.2. (Continued)

Activity	Entering Variable	Exiting Variable	Objective Coefficient Value ($/ton)	New Value of Variable (100,000 tons)	Objective Function Value (100,000 $)
x_6, landfill			−100.000	2.55500	−247.36700
	y_{s6}	y_9	−9.020	2.45280	−10.37330
	y_{s2}	y_1	−5.075	2.20752	0.02555
	y_5^a	y_5^b	−4.971	2.20752	−0.25550
	y_{s4}	y_2	−4.627	2.15846	0.73329
	y_{s8}	y_4	−1.59	0.68934	7.28872
	y_1	y_3	−0.634	0.67554	7.94742
	y_7^b	y_7^a	0.000	0.67554	8.37602
			$\boxed{4.070}$	0.67554	11.12550
	y_8	y_9	67.890	0.67554	54.23860
	y_6^b	y_6^a	91.161	0.66544	69.72420
			100.00	0.66544	75.60570
x_7, fuel			−100.000	1.66075	−150.65900
	y_{s4}	y_3	−29.324	1.63298	−33.28840
	y_{s2}	y_2	−7.453	1.48292	−2.43149
			$\boxed{-1.590}$	1.48292	11.12550
	y_1	y_3	3.110	1.48292	18.10180
	y_{s8}	y_4	4.070	1.46913	19.50560
			100.00	0.00000	19.50560

The second activity represents the tons of solid waste hauled from the area to the central plant. The corresponding cost per ton can range between $0.139 and $5.447 for a level of the activity of $x_2 = 233,016$ tons. When the cost rises above $5.447 per ton, the hauling of the solid waste must still be carried out and no significant decrease in the activity level is realized. The total cost rises at a constant rate slightly smaller than in the previous interval.

Hand-sorting of solid waste at the central plant is convenient even if no profit is made from it. In fact, at a cost of $3.934 per ton, the optimal level of the activity would be $x_3 = 5,110$ tons. This level would rise to 8,176 tons for a net revenue between $0.883 and $6.191 per ton. Above $6.191, the optimal level of the activity would jump to 28,389 tons. Above $8.738, all the possible hand-sorting of the solid waste would take place with $x_3 = 30,660$ tons.

Shredding shows a stable pattern regardless of the value of the corresponding objective function coefficient. No less than $x_4 = 224,840$ tons are shredded even when this activity costs $100 per ton. This behavior is related to the profitable fuel recovery activity whose level, x_7, depends on that of the shredding operation.

Magnetic separation also depends on shredding and remains constant in a range of net revenue from $1.223 to $100 per ton!

Landfilling is a necessity and, therefore, its level does not change for any level of cost.

The production of fuel drives, to a large extent, the solid waste management program. It would be convenient to produce $x_7 = 148,292$ tons even if the central plant had to pay $3.11 per ton rather than receive revenue for it. In that case, the total cost of the plan would be $1,810,180, while the total cost of transferring and landfilling 255,500 tons at the price of $7.40 per ton (3.33 for hauling and 4.07 for landfilling) is $1,890,700. Of course, this scenario is rather implausible and is suggested only to reinforce the idea of a waste management program. At a net revenue of $7.453 per ton, a not improbable price, the fuel recovery activity almost pays for the entire solid waste management program, which now costs only $243,149.

The model discussed in this chapter is a very simplified example of a solid waste management program. A realistic specification would include many more activities and constraints. The essential features of the problem and its analysis, however, would not be drastically different from those presented here.

Chapter 17

The Choice of Techniques in a Farm Production Model

17.1 Introduction

When introducing the farmer's problem in section 1.4, we listed five categories of activities a farmer must consider to insure the economic success of an enterprise:

1. Crop, animal, and processing activities.
2. Associated operations and their scheduling.
3. Capital, machinery, and equipment service flows.
4. Capital, machinery and equipment stock.
5. Other inputs such as fertilizer, animal feed, labor, water, etc.

These five categories are related by an intricate web of decisions requiring a large amount of information. But even when the information is available, the process of decision making is far from over. Indeed, it just begins with the crucial decision of how to organize the available information.

There are many ways to formulate a farm production plan. In chapter 3 we offered some examples of farm plans for the purpose of introducing the reader to the scope and structure of linear programming but without any pretense of realism. In order to achieve realism, or at least a satisfactory degree of practicality, the linear programming model must be as flexible as possible to encompass a large class of technical and economic scenarios.

255

Table 17.1. The Structure of a General Farm Production Model

Constraints	Production Activities	Production Techniques	Stock Activities Available	Stock Activities Purchased
Supply of fixed inputs (land)	Demand of fixed inputs (land)			
Required operations	Demand of operations	Supply of operations		
Service flow of stock activities		Demand of capital, machinery & equipment service flow	Supply of capital, machinery & equipment service flow from available stock	Supply of capital, machinery & equipment service flow from purchased stock
Supply of available stock			Demand of capital, machinery & equipment stock	Supply of purchased capital, machinery & equipment stock
Supply of other inputs	Demand & supply of other inputs	Demand of other inputs	Demand of other inputs	Demand of other inputs

When translated in terms of model formulation, flexibility means articulation of the linear programming activities into their elementary components of commodity and time dimensions. Problem 7 of section 3.8 is an example of such an articulation for a single activity. At the overall farm level, flexibility of the production plan requires giving special attention to the specification of the relations involving capital, machinery, and equipment. These relations constitute the most important component of a farm plan, since they frame the choice of techniques, the distinction between fixed and variable costs, and the farm investment schedule.

The traditional way to determine the farm demand for capital, machinery, and equipment is to preliminarily select the production techniques and the associated production coefficients. This is also the most restrictive and inflexible way. In such cases, the resulting farm plan is only conditionally optimal: no optimization with respect to techniques is considered. Clearly, the optimization with respect to production techniques cannot be conducted independently of the overall optimization process, since a production technique is optimal only in relation to the final objective of the farmer. Without sufficient knowledge of the entire production structure and the relations among all types of resources, the decision about the distribution of capital's fixed costs cannot be made in an optimal way.

The formulation of a model that also considers the choice of the optimal production techniques may appear rather complex at first, but it pays off in the end. As usual, a simple way to approach the formulation of the model is through an example. We begin with the description of the example's general structure as given in table 17.1.

The activities are grouped into three categories. Production activities include, for example, wheat, corn, alfalfa, processing, and buying and selling activities. The production techniques describe the admissible combinations of machinery and equipment for executing required cultivation practices (operations) during the various periods. These activities, in particular, define the choice of techniques. They include, for example, the combination of a 50-horsepower tractor with a given plough for executing the wheat soil preparation in a specified period. The stock activities describe the available and the needed (purchasable) stock of machinery and equipment to carry out the required operations. These activities should be associated with variables defined in integer values, but we will beg the readers' indulgence and overlook this aspect of the problem in order to maintain the illustration within a purely linear programming framework.

The constraints in table 17.1 are grouped into five categories. The first group of fixed inputs includes, for example, land and upper limits for the level of activities. The second group defines the required cultivation practices in the various periods. The third group specifies the admissible service flows of machinery and equipment. The fourth group states the stock supply of machinery and equipment, while the fifth group includes a variety of other inputs such as labor, operating capital, etc. Except for the last group of constraints, the flow of information imposes a block diagonal structure upon the matrix of technical coefficients, which is typical of well-articulated linear programming problems.

17.2   A Numerical Farm Production Model

The farm model we are about to specify exhibits 35 constraints and 70 activities (including slack variables). These dimensions prevent the usefulness of writing down the model explicitly in both its primal and dual forms. Rather, we present the detail information in table 17.2, which closely follows the structure of table 17.1. The farm production scenario and the magnitude of the parameters involved are entirely hypothetical. With this example we wish to emphasize the flexibility of linear programming and the importance of articulating a realistic flow of information to obtain a wide range of guidelines for decision making in a farm production setting with special reference to the choice of techniques.

The hypothetical farm that is the object of the LP analysis has 3,000 acres of land divided into two soil qualities. The production cycle is assumed to be a year, which is divided into three periods for accommodating the scheduling of agricultural practices required by wheat, corn, and alfalfa. The agricultural operations (soil preparation, cultivation, harvesting, etc.) can be performed with a variety of machinery and equipment, partially already available and partially purchasable, as needed. The final selection of machinery and equipment combinations constitutes the *choice of techniques,* the principal aspect we wish to emphasize in this LP model. Techniques require a service flow of machinery and equipment, which in turn requires a stock of the same inputs.

The production and processing activities envisioned on the farm include wheat and alfalfa, which can be grown on either soil A or soil B; corn, which is confined to soil A; the processing of alfalfa into pellets; and the buying and selling of corn and alfalfa pellets. The remaining activities are divided into flow activities (choice of techniques) and stock activities (already available and purchasable).

The detailed definition of the activities is given as follows:

Production activities

x_1 ≡ wheat cultivated on type A soil
x_2 ≡ wheat cultivated on type B soil

Table 17.2. Initial Tableau of the Farm Production Model

Constraint	Unit	Level	Type	x_1	x_2	x_3	x_4	x_5	x_6	x_7	x_8	x_9
1. b_1	acre	2500	$\geq$	1		1	1					
2. b_2	"	500	$\geq$		1			1				
3. b_3	hour	0	$\geq$			3						
4. b_4	"	0	$\geq$				4	3				
5. b_5	"	0	$\geq$	40	55							
6. b_6	"	0	$\geq$	2								
7. b_7	"	0	$\geq$		2							
8. b_8	"	0	$\geq$				20					
9. b_9	"	0	$\geq$									
10. b_{10}	machine	0	$\geq$									
11. b_{11}	"	0	$\geq$									
12. b_{12}	"	0	$\geq$									
13. b_{13}	"	0	$\geq$									
14. b_{14}	"	0	$\geq$									
15. b_{15}	"	0	$\geq$									
16. b_{16}	"	0	$\geq$									
17. b_{17}	"	0	$\geq$									
18. b_{18}	"	0	$\geq$									
19. b_{19}	"	0	$\geq$									
20. b_{20}	"	0	$\geq$									
21. b_{21}	"	0	$\geq$									
22. b_{22}	"	0	$\geq$									
23. b_{23}	"	0	$\geq$									
24. b_{24}	"	0	$\geq$									
25. b_{25}	"	0	$\geq$									
26. b_{26}	"	0	$\geq$									
27. b_{27}	"	0	$\geq$									
28. b_{28}	"	0	$\geq$									
29. b_{29}	"	0	$\geq$									
30. b_{30}	cwt	0	$=$			40			-1	1	-1	
31. b_{31}	"	0	$=$				400	400	-40			
32. b_{32}	thou. hours	25	$\geq$	a	a	b	c	c	d			
33. b_{33}	thou. \$	0	$\geq$	-0.5	-0.5	0.6	0.5	0.52	0.4	e	f	g
34. b_{34}	cwt	3000	$\geq$						10			
35. b_{35}	"	0	$=$						10			-1
Objective function		max		4	4.9	-1.5	-1.25	-1.1	-1	h	0.2	0.3

$a = 0.005,$ $b = 0.01,$ $c = 0.004,$ $d = 0.0001,$ $e = 0.088,$ $f = -0.03$

Table 17.2. (Continued)

	x_{10}	x_{11}	x_{12}	x_{13}	x_{14}	x_{15}	x_{16}	x_{17}	x_{18}	x_{19}	x_{20}	x_{21}	x_{22}	x_{23}	x_{24}
							Production Techniques								
1. b_1															
2. b_2															
3. b_3	-360	-340	-440	-420											
4. b_4					-450	-575									
5. b_5							i	j							
6. b_6									-350	-450					
7. b_7											-450	-400			
8. b_8													k	l	-480
9. b_9		340		420											-480
10. b_{10}	1	1			1										
11. b_{11}			1	1		1									
12. b_{12}	1		1												
13. b_{13}		1		1											
14. b_{14}					1	1									
15. b_{15}							1								
16. b_{16}								1							
17. b_{17}									1		1				
18. b_{18}										1		1			1
19. b_{19}									1	1					
20. b_{20}											1	1			
21. b_{21}													1		
22. b_{22}														1	
23. b_{23}													1	1	
24. b_{24}															1
25. b_{25}															
26. b_{26}															
27. b_{27}															
28. b_{28}															
29. b_{29}															
30. b_{30}															
31. b_{31}															
32. b_{32}															
33. b_{33}	0.84	0.88	1.08	1.12	0.8	1.04	2.7	1.6	1.26	1.62	1.32	1.68	3.06	3.96	1.9
34. b_{34}															
35. b_{35}															
O.F.	-2.1	-2.2	-2.7	-2.8	-2	-2.6	-6.75	-9	-3.15	-4.05	-3.3	-4.2	-7.65	-9.5	-4.75

$g = -0.05$, $h = -0.22$, $i = -9{,}000$, $j = -3{,}100$, $k = -4{,}800$, $l = -2{,}250$

Table 17.2. (Continued)

| | Stock Activities | | | | | | | | | | | | | |
| | Available | | | | | Purchased | | | | | | | | |
	x_{25}	x_{26}	x_{27}	x_{28}	x_{29}	x_{30}	x_{31}	x_{32}	x_{33}	x_{34}	x_{35}	x_{36}	x_{37}	x_{38}
1. b_1														
2. b_2														
3. b_3														
4. b_4														
5. b_5														
6. b_6														
7. b_7														
8. b_8														
9. b_9														
10. b_{10}	-1						-1							
11. b_{11}		-1						-1						
12. b_{12}									-1					
13. b_{13}										-1				
14. b_{14}			-1								-1			
15. b_{15}				-1								-1		
16. b_{16}													-1	
17. b_{17}	-1					-1								
18. b_{18}							-1							
19. b_{19}		-1						-1						
20. b_{20}									-1					
21. b_{21}			-1							-1				
22. b_{22}											-1			
23. b_{23}												-1		
24. b_{24}				-1										-1
25. b_{25}	1													
26. b_{26}		1												
27. b_{27}			1											
28. b_{28}				1										
29. b_{29}					1									
30. b_{30}														
31. b_{31}														
32. b_{32}														
33. b_{33}	4	0.8	1.2	14	11	2.8	4.2	4.2	14.2	27	14	10		
34. b_{34}														
35. b_{35}														
O.F.	-10	-2	-3	-30	-10	-15	-20	-4	-3	-6	-40	-50	-10	-1

$x_3 \equiv$ corn cultivated on type A soil
$x_4 \equiv$ alfalfa cultivated on type A soil
$x_5 \equiv$ alfalfa cultivated on type B soil
$x_6 \equiv$ processing of alfalfa into pellets
$x_7 \equiv$ corn purchasing
$x_8 \equiv$ corn selling
$x_9 \equiv$ pellet selling

Production techniques

period 1, corn operations:
$x_{10} \equiv$ 50-HP tractor combined with type A equipment
$x_{11} \equiv$ 50-HP tractor combined with type B equipment
$x_{12} \equiv$ 75-HP tractor with a type A equipment
$x_{13} \equiv$ 75-HP tractor with type B equipment

period 1, alfalfa operations:
$x_{14} \equiv$ 50-HP tractor with type C equipment
$x_{15} \equiv$ 75-HP tractor with type C equipment

period 2, wheat harvest operations:
$x_{16} \equiv$ SZ combine
$x_{17} \equiv$ E combine

period 3, wheat soil preparation:
$x_{18} \equiv$ 50-HP tractor with type A equipment on type A soil
$x_{19} \equiv$ 75-HP tractor with type A equipment on type A soil
$x_{20} \equiv$ 50-HP tractor with type A equipment on type B soil
$x_{21} \equiv$ 75-HP tractor with type A equipment on type B soil

period 3, corn harvest operations:
$x_{22} \equiv$ SZ combine with type E equipment
$x_{23} \equiv$ E combine with type E equipment
$x_{24} \equiv$ 75-HP tractor with type Z equipment

Machinery and equipment stock activities

available stock:
$x_{25} \equiv$ 50-HP tractors
$x_{26} \equiv$ type A equipment
$x_{27} \equiv$ type C equipment
$x_{28} \equiv$ SZ combines
$x_{29} \equiv$ type A equipment

purchased stock:
$x_{30} \equiv$ 50-HP tractor
$x_{31} \equiv$ 75-HP tractor
$x_{32} \equiv$ type A equipment

$x_{33} \equiv$ type B equipment
$x_{34} \equiv$ type C equipment
$x_{35} \equiv$ SZ combines
$x_{36} \equiv$ E combines
$x_{37} \equiv$ type E equipment
$x_{38} \equiv$ type Z equipment

The constraints are specified as the following five groups:

Fixed inputs

$b_1 \equiv$ land—type A soil
$b_2 \equiv$ land—type B soil

Required operations

period 1

$b_3 \equiv$ corn cultivation
$b_4 \equiv$ alfalfa cultivation

period 2

$b_5 \equiv$ wheat harvest

period 3

$b_6 \equiv$ type A soil preparation for wheat
$b_7 \equiv$ type B soil preparation for wheat
$b_8 \equiv$ corn harvest with combine
$b_9 \equiv$ corn harvest with 75-HP tractor and type Z equipment

Machinery and equipment service flow

period 1

$b_{10} \equiv$ 50-HP tractors
$b_{11} \equiv$ 75-HP tractors
$b_{12} \equiv$ type A equipment
$b_{13} \equiv$ type B equipment
$b_{14} \equiv$ type C equipment

period 2

$b_{15} \equiv$ SZ combines
$b_{16} \equiv$ E combines

period 3

$b_{17} \equiv$ 50-HP tractors
$b_{18} \equiv$ 75-tractors
$b_{19} \equiv$ type A equipment
$b_{20} \equiv$ type B equipment
$b_{21} \equiv$ SZ combines
$b_{22} \equiv$ E combines

$b_{23} \equiv$ type E equipment
$b_{24} \equiv$ type Z equipment

Machinery and equipment stock

available

$b_{25} \equiv$ 50-HP tractors
$b_{26} \equiv$ type A equipment
$b_{27} \equiv$ type C equipment
$b_{28} \equiv$ SZ combines
$b_{29} \equiv$ type Z equipment

Other inputs

$b_{30} \equiv$ corn material balance
$b_{31} \equiv$ alfalfa material balance
$b_{32} \equiv$ labor
$b_{33} \equiv$ operating capital
$b_{34} \equiv$ pellets plant capacity
$b_{35} \equiv$ pellets material balance

The specification of an articulated linear program requires a considerable amount of information even for a stylized farm production plan as illustrated by this example. When that requirement is met, however, the LP model can offer interesting insights on the available strategies. It is not obvious, in fact, how to choose among wheat, alfalfa, and corn for maximum profit, and even less transparent how to select the techniques to guarantee that goal.

The optimal primal and dual solutions are given in tables 17.3 and 17.4, respectively. The optimal value of the objective function is $17,031,898.40.

Consider table 17.3, first. The "B" mark associated with the variable names indicates that the corresponding activity is in the primal basis, while the "NB" mark means nonbasic. Land of type A soil is allocated among wheat (x_1), corn (x_3), and alfalfa (x_4). Land of type B soil goes to wheat (x_2). The required operations are to be accomplished with combinations of 50-HP tractors $(x_{10}, x_{14}, x_{18},$ and $x_{20})$, SZ combines $(x_{16}$ and $x_{22})$ and type A, B, C, and E equipment. The available stock of 50-HP tractors (x_{25}), SZ combines (x_{28}), and type A equipment (x_{26}) is fully utilized and, indeed, it is not sufficient to perform all the required agricultural practices. For this reason, 7 additional 50-HP tractors (x_{30}), 5 SZ combines (x_{35}), 10 items of type A equipment (x_{32}), 2 of type B (x_{33}), and 8 of type E (x_{37}) must be purchased.

Notice that the primal solution is degenerate in four of its components marked with a single asterisk (*) at the zero level and the zero opportunity cost. Three of those components relate to 75-HP tractors $(x_{13}, x_{24},$ and

x_{31}), while the fourth one is associated with an E combine (x_{23}). This means that this farm problem might have multiple dual solutions if a pivot for the dual algorithm can be found. Furthermore, the nonbasic activity associated with x_{19} and relating to a combination of 75-HP tractors and type A equipment shows a zero opportunity cost (marked by a double asterisk **), indicating that this same problem also has multiple primal solutions. The search for multiple solutions was not carried out, but the available information allows us to conclude that a combination of 75-HP tractors and various equipment could also be used as an alternative to the techniques based upon 50-HP tractors.

The sensitivity analysis of the objective function coefficients (called range analysis in table 17.3) shows that the optimal solution is stable, since the range of variation of each price coefficient is rather wide. The existence of multiple optimal solutions in association with the wide ranges of price coefficients allows us to say that the set of such a solution is also stable. It is rather unfortunate that commercial computer programs do not incorporate the option of explicitly finding at least some of the optimal solutions of a given linear programming problem. This not easily explainable omission confines much useful information to the grip of the software for the large majority of LP users. Since demand creates much of its own supply, it is imperative, therefore, that LP users become aware of the opportunity for retrieving the missing information and begin requesting that computer programs be equipped with the capability of computing multiple optimal solutions when they exist.

Consider, now, the dual solution presented in table 17.4. It is also degenerate in two components (y_{16} and y_{18}), a fact that increases the importance of multiple solutions. The shadow price of land of type A soil is \$1,820 (price coefficients in the primal objective function were expressed in thousands of dollars), while the land of type B soil commands a shadow price of \$2,699. The required agricultural operations have a shadow price that varies from \$4 to \$59 per hour, as indicated by the dual variables associated with constraints b_3 through b_9. The shadow price of the service flow of various machinery and equipment combinations ranges in the thousands of dollars, since the units of measurement refer to the total hours of service flow obtainable from the combination in a given period. Hence, the shadow price of b_{21} (which refers to the service flow of a SZ combine in period 3) is \$40,000, the service flow of b_{10} (a 50-HP tractor in period 1) is \$15,000, while that of b_{11} (a 75-HP tractor in the same period) is \$20,000. The shadow prices of the equipment's service flow is much lower, ranging from \$3,000 of b_{20} (type B) to \$10,000 of b_{23} (type E).

Table 17.3. Primal Solution and Range Analysis of the Objective Function

Activity	Level	Opportunity Cost	Objective Coefficient	Allowable Increase	Allowable Decrease
x_1 B	470.00	0.000	4.00	0.684	0.910
x_2 B	500.00	0.000	4.90	infinity	0.684
x_3 B	2,000.00	0.000	−1.50	1.820	2.132
x_4 B	30.00	0.000	−1.25	infinity	0.684
x_5 NB	0.00	0.684	−1.10	0.684	infinity
x_6 B	300.00	0.000	−1.00	infinity	1.262
x_7 NB	0.00	0.020	0.22	0.020	infinity
x_8 B	79,700.00	0.000	0.20	0.020	0.053
x_9 B	3,000.00	0.000	0.30	infinity	0.126
Techniques					
x_{10} B	16.67	0.000	−2.10	1.557	0.745
x_{11} NB	0.00	1.743	−2.20	1.743	infinity
x_{12} NB	0.00	0.911	−2.70	0.911	infinity
x_{13} B	*0.00	*0.000	−2.80	6.045	1.817
x_{14} B	0.27	0.000	−2.00	20.000	0.035
x_{15} NB	0.00	0.044	−2.60	0.044	infinity
x_{16} B	5.14	0.000	−6.75	6.750	1.687
x_{17} NB	0.00	0.600	−3.00	0.600	infinity
x_{18} B	2.69	0.000	−3.15	3.150	0.000
x_{19} NB	0.00	**0.000	−4.05	0.000	infinity
x_{20} B	2.22	0.000	−3.30	6.300	1.800
x_{21} NB	0.00	1.600	−4.20	1.600	infinity
x_{22} B	8.33	0.000	−7.65	16.050	69.088
x_{23} B	*0.00	*0.000	−9.50	42.477	7.523
x_{24} B	*0.00	*0.000	−4.75	6.909	3.091
Stock					
x_{25} B	10.00	0.000	−10.00	infinity	5.000
x_{26} B	7.00	0.000	−2.00	infinity	2.00
x_{27} B	0.27	0.000	−3.00	3.000	0.160
x_{28} B	3.00	0.000	−30.00	infinity	10.000
x_{29} NB	0.00	6.909	−10.00	6.909	infinity
x_{30} B	6.93	0.000	−15.00	1.557	0.035
x_{31} B	*0.00	*0.000	−20.00	0.044	1.817
x_{32} B	9.67	0.000	−4.00	1.557	4.100
x_{33} B	2.22	0.000	−3.00	1.743	153.969
x_{34} NB	0.00	3.000	−6.00	3.000	infinity
x_{35} B	5.33	0.000	−40.00	10.000	69.088
x_{36} NB	0.00	42.476	−50.00	42.477	infinity
x_{37} B	8.33	0.000	−10.00	10.000	14.162
x_{38} B	0.00	11.909	−15.00	11.909	infinity

Table 17.4. Dual Solution and Range Analysis of the Right-Hand Side

Constraint	Dual Price	Slack Variable	Original Level	Allowable Increase	Allowable Decrease
b_1 NB	1.820	0.000	2,500	244	235
b_2 NB	2.699	0.000	500	216	467
Required Operations					
b_3 NB	0.059	0.000	0	2,496	25,193
b_4 NB	0.044	0.000	0	120	1,680
b_5 NB	0.009	0.000	0	46,300	28,700
b_6 NB	0.009	0.000	0	940	4,208
b_7 NB	0.014	0.000	0	1,000	5,411
b_8 NB	0.012	0.000	0	15,306	189,557
b_9 NB	0.004	0.000	0	933	0
Service Flow					
b_{10} NB	15.000	0.000	0	7	88
b_{11} NB	20.000	0.000	0	0	112
b_{12} NB	4.000	0.000	0	10	441
b_{13} B	0.000	2.222	0	infinity	2
b_{14} NB	3.000	0.000	0	0	4
b_{15} B	0.000	3.189	0	infinity	3
b_{16} B	*0.000	*0.000	0	infinity	0
b_{17} B	0.000	12.025	0	infinity	12
b_{18} B	*0.000	*0.000	0	infinity	0
b_{19} NB	0.000	13.980	0	infinity	14
b_{20} NB	3.000	0.000	0	2	294
b_{21} NB	40.000	0.000	0	3	87
b_{22} NB	7.523	0.000	0	7	0
b_{23} NB	10.000	0.000	0	8	88
b_{24} NB	3.091	0.000	0	2	0
Available Stock					
b_{25} NB	5.000	0.000	10	7	10
b_{26} NB	2.000	0.000	7	10	7
b_{27} B	0.000	3.733	4	infinity	4
b_{28} NB	10.000	0.000	3	5	3
b_{29} B	0.000	4.000	4	infinity	4
Other Inputs					
b_{30} NB	−0.200	0.000	0	41,150	infinity
b_{31} NB	−0.012	0.000	0	156,667	12,000
b_{32} NB	426.392	0.000	25	2	2
b_{33} B	0.000	1,234.429	0	infinity	1,234
b_{34} NB	0.126	0.000	3,000	42,000	3,000
b_{35} NB	−0.300	0.000	0	3,000	infinity

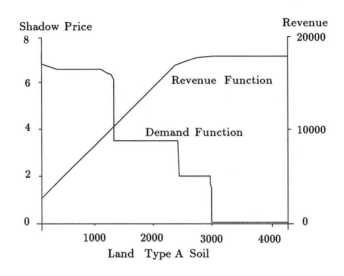

Figure 17.1. Parametric analysis of land, type A soil.

The shadow price of the available stock of machinery and equipment seem to reflect the used aspect of this capital. A 50-HP tractor (b_{25}) is worth \$5,000, a used SZ combine (b_{29}) \$10,000, and a used item of type A equipment (b_{26}) \$2,000. In this environment, labor (b_{32}) exhibits a value marginal product of \$0.43 per hour. The sensitivity analysis of the right-hand-side coefficients indicate that the set of optimal solutions is rather stable, since the upper and lower bounds of the range interval are always rather far apart.

17.3 The Derived Demands for Land and Labor

From an economic and entrepreneurial viewpoint, the evaluation of the information contained in a linear programming model is never complete without the parametric analysis of the principal resources. The adjustment of these resources, which is in principle under the control of the farmer, may be a crucial step toward an improved management strategy. In this example, we have selected land and labor to illustrate the type of information to be derived from their parametric analysis. Let us examine first the land of type A soil whose parametric information is summarized in figure 17.1.

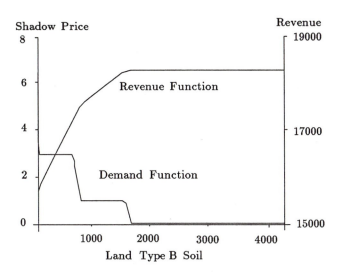

Figure 17.2. Parametric analysis of land, type B soil.

The initial availability of this type of land is 2,500 acres with an associated shadow price of $1,820 per acre. As the availability is varied from 0 to 10,000 acres, the corresponding shadow price drops from about $7,190 to $0.00. Actually, the zero shadow price is achieved at a level of 2,765 acres, which is very close to the initial availability of type A land. Correspondingly, the revenue function exhibits its distinct concave curvature.

The parametric analysis of land of type B quality shows a similar pattern with noticeable differences (figure 17.2). At the initial endowment level of 500 acres, type B soil has a shadow price higher than that of type A. However, it never reaches the productivity levels of type A soil except in the vicinity of zero. The same shadow price drops to about $800 per acre when the availability is expanded to about 770 acres. The comparison of the two parametric analyses, therefore, has shown that, in spite of the initial information giving that type B soil as being the more productive, the land of type A quality is in the end more valuable over a wide interval.

The parametric analysis of labor (figure 17.3) indicates that, given the choice of production activities and techniques revealed by the optimal solution, its productivity is relatively stable within a supply range that stretches from 7,000 to 27,000 man-hours. This behavior corresponds to a

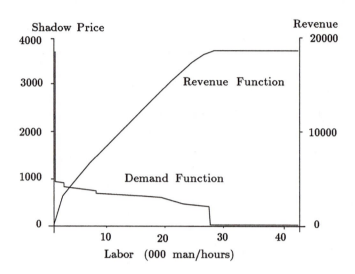

Figure 17.3. Parametric analysis of labor.

rather elastic demand function, where the elasticity index is defined as the percentage change in labor quantity over the percentage change in labor shadow price.

Chapter 18

Cattle Ranch Management

18.1 A Static Model of Ranch Management

Linear programming has been used for many years to help make management decisions in beef feedlot and dairy cattle operations. In this chapter we discuss a LP model applied to range cattle ranching. The main problem of this type of farm operation is the decision about the stock of animals versus the available quantities of feed during the year. Cattle ranching is a dynamic enterprise requiring a sequence of decisions overlapping several production cycles. In order to keep the problem relatively simple, however, we begin with a static specification of the problem and turn to a multistage formulation in a further section.

A second motivation for discussing this problem is that it is suitable for illustrating the adaptive process of formulating a reasonable plan. We must emphasize that the specification of a linear programming model is never a one-shot deal. Often, what seem plausible and intuitive relations turn out to be too restrictive and require modification.

The model discussed here is based on a hypothetical beef cattle ranch on annual range in San Luis Obispo county in California. It was originally formulated by Weitkamp, Clawson, Center, and Williams of the Cooperative Extension Service and the University of California at Davis. The year is divided into five periods or seasons based upon the changing pattern of forage availability and animal feed requirements as reported in table 18.1.

Table 18.2 gives daily and seasonal nutrient requirements of the cattle units in pounds of dry matter and crude protein. Dry matter (a measure of feed quantity) and crude protein (a measure of feed quality) are used in this model because they are readily available. More correct indicators of nutrient content such as total digestible nutrients, metabolizable energy, digestible energy, and digestible crude protein can easily be substituted

271

Table 18.1. Seasons, Range Forage, and Cow Status

Season	Dates	Days	Description
1	Nov. 21–Feb. 15	87	Inadequate green forage, cows lactating
2	Feb. 16–May 15	89	Adequate green forage, cows late-lactating
3	May 16–Aug. 1	78	Dry forage, calves weaned
4	Aug. 2–Sept. 20	50	Dry, leached forage, late gestation
5	Sept. 21–Nov. 20	61	Dry, leached forage, supplement, cows calving

when available. The assumption is that the ranch is stocked with a herd of cattle with fixed composition: cows, calves (90 percent of calf crop), heifers (20 percent replacement), and bulls (1/20 of cows). The animal products include calves, native (home-bred) steers, and purchased steers.

The ranch is divided into several land units depending on terrain and soil conditions. There are 3,000 acres of unimproved range that will be grazed "as is"; 1,000 acres of improvable range that can be treated with three management alternatives: graze as is, fertilize, and seed with legumes and fertilize; 1,000 acres of brushland that can either be grazed as is or cleared; 75 acres of irrigated pasture that can either be grazed all five seasons, or grazed in seasons 1, 3, 4, and 5 while producing hay in season 2; 25 acres are cultivated with alfalfa for hay that can either be fed to the ranch cattle or sold. Hay production is a typical farming operation on many ranches and is included in the model as a product alternative on irrigated pasture and alfalfa land. Supplemental feeding of brood cows, which is necessary for satisfactory calf production, is included as two purchase options in the form of liquid protein supplement and oat hay. Ranch-produced pasture and alfalfa hay are also available for supplementing the cows' diet.

A detailed definition of the activities and constraints is as follows:

Activities

land management alternatives (acres):

x_1 ≡ irrigated pasture, graze in seasons 1, 3, 4, 5, hay in season 2
x_2 ≡ irrigated pasture, graze in all five seasons
x_3 ≡ unimproved range, graze as is
x_4 ≡ improvable range, graze as is
x_5 ≡ improvable range, fertilize
x_6 ≡ improvable range, seed and fertilize
x_7 ≡ brushland, graze as is

Table 18.2. Cattle Requirements for Dry Matter and Crude Protein

	Season 1		Season 2	
	Head	87	Head	89
Animal	day	days	day	days
Per cow-calf unit	Dry matter (pounds)			
Cow	21.8	1897	22.0	1958
Calf (90% calf crop)			3.0	240
Heifer (20% replacement)	18.1	315	19.8	352
Bull (1/20 cows)	28.0	122	28.0	125
Total cow-calf unit		2334		2675
Native steer	15.0	1305	18.0	1602
Purchased steer	15.0	1305	18.0	1602
Per cow-calf unit	Crude protein (pounds)			
Cow	2.0	174	2.0	178
Calf (90% calf crop)			0.4	16
Heifer (20% replacement)	1.8	31	1.9	34
Bull (1/20 cows)	3.1	13	3.1	14
Total cow-calf unit		218		242
Native steer	1.5	130	1.7	151
Purchased steer	1.5	130	1.7	151

Table 18.2. (Continued)

	Season 3		Season 4		Season 5	
	Head	78	Head	50	Head	61
Animal	day	days	day	days	day	days
Per cow-calf unit	Dry matter (pounds)					
Cow	17.6	1373	15.0	750	19.8	1208
Calf	4.0	112				
Heifer	11.0	103	12.8	128	14.5	177
Bull	28.0	109	28.0	170	28.0	85
Total cow-calf unit		1697		948		1470
Native steer	19.0	821	11.0	550	11.0	671
Purchased steer	19.0	304				
Per cow-calf unit	Crude protein (pounds)					
Cow	1.3	101	0.9	45	2.0	122
Calf	0.5	8				
Heifer	1.2	11	1.4	14	1.5	18
Bull	3.1	12	3.1	8	3.1	9
Total cow-calf unit		132		67		149
Native steer	1.8	81	1.1	55	1.1	67
Purchased steer	1.8	29				

x_8 ≡ brushland, graze after clearing
x_9 ≡ alfalfa hay

feed:

x_{10} ≡ alfalfa ranch hay, season 1, tons
x_{11} ≡ alfalfa ranch hay, season 4, tons
x_{12} ≡ alfalfa ranch hay, season 5, tons
x_{13} ≡ irrigated pasture ranch hay, season 1, tons
x_{14} ≡ irrigated pasture ranch hay, season 4, tons
x_{15} ≡ irrigated pasture ranch hay, season 5, tons
x_{16} ≡ purchased liquid protein, season 1, pounds
x_{17} ≡ purchased liquid protein, season 4, pounds
x_{18} ≡ purchased liquid protein, season 5, pounds
x_{19} ≡ purchased oat hay, season 1, tons
x_{20} ≡ purchased oat hay, season 2, tons
x_{21} ≡ purchased oat hay, season 4, tons

products:

x_{22} ≡ alfalfa hay, tons
x_{23} ≡ pasture hay, tons
x_{24} ≡ cow-calves
x_{25} ≡ native steers
x_{26} ≡ purchased steers

Constraints

b_1 ≡ irrigated pasture, acres
b_2 ≡ unimproved range, acres
b_3 ≡ improvable range, acres
b_4 ≡ brushland, acres
b_5 ≡ alfalfa, acres
b_6 ≡ capital, dollars
b_7 ≡ pasture hay, dry matter, pounds
b_8 ≡ alfalfa hay, dry matter, pounds
b_9 ≡ forage dry matter, season 1, pounds
b_{10} ≡ forage dry matter, season 2, pounds
b_{11} ≡ forage dry matter, season 3 pounds
b_{12} ≡ forage dry matter, season 4, pounds
b_{13} ≡ forage dry matter, season 5, pounds
b_{14} ≡ forage crude protein, season 1, pounds
b_{15} ≡ forage crude protein, season 2, pounds

Table 18.3. Initial Tableau of the Cattle Ranch Management Problem

Con-straint	Unit	Level	Type	Land management strategies ($x_1 - x_9$)							
				x_1	x_2	x_3	x_4	x_5	x_6	x_7	x_8
b_1	acre	75	≥	1	1						
b_2	acre	3000	≥			1					
b_3	acre	1000	≥				1	1	1		
b_4	acre	1000	≥							1	1
b_5	acre	25	≥								
b_6	tho\$	100	≥	120	80	0.5	0.5	11.9	10.4	0.2	1.6
b_7	lb	0	≤	3520							
b_8	lb	0	≤								
b_9	lb	0	=	500	500	100	250	450	400	2	75
b_{10}	lb	0	=		3520	128	320	576	512	3	96
b_{11}	lb	0	=	2880	2880	72	180	324	288	2	54
b_{12}	lb	0	=	1400	1400	40	100	180	160	1	30
b_{13}	lb	0	=	2200	2200	60	150	270	240	2	45
b_{14}	lb	0	≤	60	60	10	25	63	48	0.2	8
b_{15}	lb	0	≤		634	19	48	92	92	0.5	14
b_{16}	lb	0	≤	576	576	4	13	26	35	0.2	3
b_{17}	lb	0	≤	252	252	2	6	7	16	0.1	1
b_{18}	lb	0	≤	396	396	2	8	8	24	0.2	1
b_{19}	head	600	≥								
b_{20}	head	260	≥								
obj. function		0	=	-120	-80	-0.5	-0.5	-11.9	-10.4	-0.2	-1.6

$b_{16} \equiv$ forage crude protein, season 3, pounds
$b_{17} \equiv$ forage crude protein, season 4, pounds
$b_{18} \equiv$ forage crude protein, season 5, pounds
$b_{19} \equiv$ cows, heads
$b_{20} \equiv$ native steers, heads

The initial tableau of the cattle ranch management problem is presented in table 18.3. The estimated amount of forage available on the various soils corresponds to a "fair" year, evaluated as 67 percent production of an "average" year. One critical feature of a linear programming model is whether constraints should be specified as inequalities even when equalities would seem to make intuitive sense. Consider the dry matter and the crude protein constraints of table 18.3. Their meaning is that the forage produced must be equal to the forage consumed or sold on the market. It would seem sensible to specify these relations as equalities, since wasting forage is costly. On the other hand, equalities represent tight (binding) constraints for which we may have to pay a price. For these reasons, the initial specification shows that the dry matter constraints are equations while the crude protein constraints are inequalities.

Table 18.3. (Continued)

Con-straint	x_9	x_{10}	x_{11}	x_{12}	x_{13}	x_{14}	x_{15}	x_{16}	x_{17}	x_{18}
				Feed $(x_{10} - x_{21})$						
b_1										
b_2										
b_3										
b_4										
b_5	1									
b_6	400							.044	.044	.044
b_7					-1760	-1760	-1760			
b_8	14,480	-1810	-1810	-1810						
b_9		1810			1760			1		
b_{10}										
b_{11}										
b_{12}			1810			1760			1	
b_{13}				1810			1760			1
b_{14}		350			320			0.2		
b_{15}										
b_{16}										
b_{17}			350			320			0.2	
b_{18}				350			320			0.2
b_{19}										
b_{20}										
o.f.	-400	0	0	0	0	0	0	-.044	-.044	-.044

The optimal primal solution is presented in table 18.4. The value of the objective function is $25,550.50 for an average return of $5.01 per acre. This level of performance is obtained by allocating the irrigated pasture between x_1 and x_2, that is, grazing it for four seasons and grazing all five seasons, with a surplus of 4.13 acres. This is an immediate indication that the model could be modified to allow the exploitation of all the available irrigated pastureland. The unimproved range is entirely utilized to support the cattle herd. The management strategy for the improvable range indicates that the seeding with fertilizer option (x_6) is to be selected. The 1,000 acres of brushland should be cleared. The 25 acres of alfalfa hay are entirely utilized for producing hay to be sold on the market without any ranch utilization (x_9). The hay produced from the irrigated pasture during season 2 will be used in season 1 (x_{13}). This seems an impossibility, except that we must consider this static model as a yearly photograph in a sequence of cycles of the same type. The solution, therefore, must be read in the sense that the hay produced in season 2 will be utilized in season 1 of the next cycle. Liquid protein and oat hay (x_{18}, x_{19}, and x_{21}) are to be purchased in abundance to supplement the diet. Finally, the marketable products of this cattle ranch are the alfalfa hay (x_{22}) and the calves from the cow-calf activity (x_{24}). No steers, either native or purchased, are profitable under the conditions of table 18.3. The dual solution is degenerate, since the opportunity cost associated with the nonbasic activity #14 is equal to zero. Also, the opportunity costs of activities #16 and #17 are very small, suggesting the possibility of nearly optimal solutions.

Table 18.3. (Continued)

Constraint	x_{19}	x_{20}	x_{21}	x_{22}	x_{23}	x_{24}	x_{25}	x_{26}
				Final products ($x_{22} - x_{26}$)				
b_1								
b_2								
b_3								
b_4								
b_5								
b_6	70	70	70					
b_7					-1760			
b_8				-1810				
b_9	1764					-2334	-1305	-1305
b_{10}		1764				-2675	-1602	-1602
b_{11}						-1697	-821	-304
b_{12}			1764			-948	-550	
b_{13}						-1470	-671	
b_{14}	184					-218	-130	-130
b_{15}		184				-242	-151	-151
b_{16}						-132	-81	-29
b_{17}			184			-67	-55	
b_{18}						-149	-67	
b_{19}						1		
b_{20}							1	
o.f.	-70	-70	-70	80	65	110	55	50

Table 18.4. Primal Solution of the Cattle Ranch Management Problem

| | | | Objective Coefficients Ranges | | |
Activity	Level	Opportunity Cost	Original Coefficient	Allowable Increase	Allowable Decrease
x_1	11.41	0.00	−120.00	40.00	29.93
x_2	59.47	0.00	−80.00	9.47	40.00
x_3	3,000.00	0.00	−0.50	infinity	2.96
x_4	0.00	0.49	−0.50	0.49	infinity
x_5	0.00	6.56	−11.90	6.56	infinity
x_6	1,000.00	0.00	−10.40	infinity	0.49
x_7	0.00	0.91	−0.20	0.91	infinity
x_8	1,000.00	0.00	−1.60	infinity	0.80
x_9	25.00	0.00	−400.00	infinity	240.00
x_{10}	0.00	8.17	0.00	8.17	infinity
x_{11}	0.00	8.17	0.00	8.17	infinity
x_{12}	0.00	5.16	0.00	5.16	infinity
x_{13}	22.81	0.00	0.00	20.00	infinity
x_{14}	*0.00	0.00	0.00	0.00	infinity
x_{15}	0.00	5.20	0.00	5.20	infinity
x_{16}	0.00	0.004	−0.044	0.004	infinity
x_{17}	0.00	0.004	−0.044	0.004	infinity
x_{18}	39,248.54	0.00	−0.044	0.071	0.003
x_{19}	112.02	0.00	−70.00	0.00	7.62
x_{20}	0.00	20.05	−70.00	20.05	infinity
x_{21}	9.36	0.00	−70.00	122.36	0.00
x_{22}	200.00	0.00	80.00	infinity	5.16
x_{23}	0.00	4.84	65.00	4.84	infinity
x_{24}	449.00	0.00	110.00	19.33	13.26
x_{25}	0.00	16.95	55.00	16.95	infinity
x_{26}	0.00	28.76	50.00	28.76	infinity

The dual variables are reported in table 18.5. The marginal productivity of the various types of land shows that unimproved land (b_2) is valued at \$2.96 per acre, improved range with the seed and fertilizer option (b_3) command \$9.86 per acre, while the brushland shadow price is \$0.80 per acre. The land devoted to the production of alfalfa hay (b_5) exhibits a marginal value of \$240 per acre. The initial availability of capital (\$50,000) is more than sufficient to cover the need of the ranch operations and thus its shadow price is zero. The shadow prices of dry matter ($b_9 - b_{13}$) and

Table 18.5. Dual Variables of the Cattle Ranch Management Problem

| | | | Right-Hand-Side Ranges | | |
Constraint	Surplus Level	Dual Variable	Current RHS	Allowable Increase	Allowable Decrease
b_1	4.13	0.00	75.00	infinity	4.13
b_2	0.00	2.96	3,000.00	216.83	2,829.17
b_3	0.00	9.86	1,000.00	916.79	657.14
b_4	0.00	0.80	1,000.00	246.41	1,000.00
b_5	0.00	240.00	25.00	25.38	25.00
b_6	10,150.39	0.00	50,000.00	infinity	10,150.39
b_7	0.00	0.040	0.00	40,145.64	197,599.37
b_8	0.00	0.044	0.00	362,000.03	infinity
b_9	0.00	−0.040	0.00	255,789.78	194,238.52
b_{10}	0.00	−0.028	0.00	40,145.64	197,599.37
b_{11}	0.00	0.061	0.00	9,764.83	35,057.87
b_{12}	0.00	−0.040	0.00	255,789.78	16,519.50
b_{13}	0.00	0.040	0.00	44,557.68	4,171.77
b_{14}	20,260.71	0.00	0.00	20,260.71	infinity
b_{15}	92,021.30	0.00	0.00	92,021.30	infinity
b_{16}	31,541.67	0.00	0.00	31,541.67	infinity
b_{17}	12,493.50	0.00	0.00	12,493.50	infinity
b_{18}	0.00	0.40	0.00	834.35	8,922.54
b_{19}	151.00	0.00	600.00	infinity	151.00
b_{20}	260.00	0.00	260.00	infinity	260.00

crude protein ($b_{14} - b_{18}$) indicate that dry matter is in short supply over crude protein. This is another indication for the need of improving the present specification and allowing a more flexible utilization of all feeding material. Dry matter is most scarce during season 3, as indicated by the corresponding shadow price of $0.061 per pound.

We have selected three crucial inputs (improved range, b_3, capital, b_6, and cows, b_{19}) for parametric analysis, as reported in table 18.6. Improved range must more than double (2,283 acres) before it requires a change of basis. At that time, the activity native steers (x_{25}) enters the profitable combination of outputs. The analysis of capital shows that the present level of revenue can be sustained with about $39,000. When capital is available at $30,000, its shadow price is about 11 percent for a revenue level of about $24,000. The comparison of marginal earnings achievable in other forms of investment indicates that $30,000 is about the maximum amount of capital to be sunk into this ranch, *ceteris paribus*. The analysis of cows suggests that a herd of about 300–350 units is a sensible strategy, given the initial conditions. At that level, the activity native steers enters the basis and contributes to a more balanced herd.

Table 18.6. Parametric Analysis of Improved Range, Capital and Cows

Input	Activity Exiting Basis	Activity Entering Basis	Input Level	Shadow Price	Objective Function Value
Improved			3,000.00	3.405	39,965.4
range	slack 7	slack 3	2,621.35	3.779	38,675.9
	x_8	x_{25}	2,282.87	7.675	37,396.9
	slack 20	slack 5	1,916.79	9.857	34,587.5
	x_{13}	x_{20}	342.86	9.857	19,073.1
	x_{16}	slack 4	0.00	10.551	15,455.5
Capital			50,000.0	0.000	25,550.5
	slack 7	x_{12}	39,849.6	0.000	25,550.5
	x_{18}	x_{10}	38,453.6	0.081	25,437.6
	x_{19}	x_{11}	30,381.2	0.114	24,518.8
	x_{21}	slack 5	29,714.8	0.114	24,443.0
	x_8	x_4	26,829.7	0.163	23,971.4
	slack 2	slack 3	22,601.8	0.180	23,211.6
	x_{13}	slack 6	19,549.0	0.547	21,540.6
	x_{22}	x_{20}	17,998.7	0.600	20,610.4
	x_6	slack 2	16,354.9	0.736	19,400.6
	x_{20}	x_{13}	8,808.9	0.876	12,791.3
	x_3	slack 4	6,880.0	0.929	10,999.4
			0.0	1.599	0.0
Cows			600	0.00	25,550.5
	slack 20	slack 5	449	0.00	25,550.5
	x_8	x_{25}	389	13.26	24,751.7
	x_{13}	slack 3	323	36.93	22,317.3
	x_2	x_{20}	79	39.43	12,690.6
	x_3	x_2	70	44.25	12,301.4
	x_{18}	x_4	55	44.92	11,615.4

Overall, the solution and the analysis associated with the initial specification of the problem as presented in table 18.3 are not entirely satisfactory. The lack of complete utilization of the 75 acres of irrigated pasture, while so much supplemental feed in the form of liquid protein and oat hay has to be purchased, casts serious doubts on the articulation of the model in its present form. To evaluate the impact the dry matter equations might have had on this outcome, we replace them by inequalities. With only this change, the new solution shows a dramatic improvement. The value of the objective function rises to $34,735.43. All the different types of land are now completely utilized. No supplemental purchase of feed is required, a fact that reduces the capital expense. Steer production is still not recommended.

The optimal combination of primal activities is as follows:

x_1	75.00	irrigated pasture, graze seasons 1, 3, 4, 5, hay season 2
x_3	3,000.00	unimproved range
x_4	1,000.00	improvable range, as is
x_8	1,000.00	brushland, cleared
x_9	25.00	alfalfa hay
x_{13}	20.18	irrigated pasture hay, supplement
x_{22}	200.00	alfalfa hay, sold
x_{23}	129.82	pasture hay, sold
x_{24}	299.00	cows

The shadow prices of the primal constraints are:

y_1	28.46	irrigated pasture
y_2	4.33	unimproved range
y_3	11.58	improvable range
y_4	2.02	brushland
y_5	240.00	alfalfa
y_7	0.037	pasture hay, dry matter
y_8	0.044	alfalfa, dry matter
y_9	0.037	forage, dry matter, season 1
y_{10}	0.009	forage, dry matter season 2

Now that all the land is put into production, this revised plan seems more acceptable than the original solution. However, a considerable imbalance still remains in the utilization of forage throughout the various seasons. This is revealed by the fact that only the dual variables associated with dry matter in seasons 1 and 2 are positive. The verification of this statement can be made by considering the slack variables (pounds) of dry matter and crude protein constraints resulting from the revised specification:

slack 9	0.00	forage, dry matter, season 1
slack 10	0.00	forage, dry matter, season 2
slack 11	158,485.98	forage, dry matter, season 3
slack 12	71,485.98	forage, dry matter, season 4
slack 13	100,373.83	forage, dry matter, season 5

slack 14	8,761.68	forage, crude protein, season 1
slack 15	46,626.17	forage, crude protein, season 2
slack 16	31,723.37	forage, crude protein, season 3
slack 17	11,862.62	forage, crude protein, season 4
slack 18	139.25	forage, crude protein, season 5

Clearly, the nonutilization of so much forage during the various seasons is not satisfactory. This imbalance is due to the static formulation of the problem, which does not allow the transfer of surplus feed from one season to the next. To model this possibility, it is necessary to conceive a multistage version of the ranch management problem.

18.2   A Multistage Model of Ranch Management

The main characteristic of multistage models is the linking of the various periods in a pseudodynamic way to allow the transfer of unused resources from one period to the next. To accomplish this objective, many new activities and constraints are required. This enlargement of the problem does not create any particular difficulty in the computational procedure. Economic problems of several thousand constraints and more than twenty thousand variables have been successfully formulated and solved. The expansion of the problem, however, becomes a burden for an expository book such as this. For this reason we will discuss a reduced formulation of the multistage version of the cattle ranch management problem.

The inadequacy of the static model was identified in the poor utilization of the feed resources during the five seasons. It was also verified that the crude protein content never placed a restriction on the nutritional requirements of the animal diet. For this reason we will disregard the corresponding constraints in the multistage specification and concentrate on modeling the dry matter availabilities. The feed sources of the ranch are, as before, irrigated pasture, unimproved range, improvable range, brushland, and alfalfa. We will retain all the corresponding activities, but only those associated with unimproved and improvable range will be specified in a multistage fashion. Even with these simplifications, the original static model with 20 constraints and 26 variables will expand into a multistage model with 47 constraints and 74 variables.

In a multistage cattle ranch management model, the forage production is entered as seasonal growth rather than as an estimate of livestock requirements. Table 18.7 reports the information necessary for implementing this model.

The recursive nature of a multistage cattle ranch model is highlighted

Table 18.7. Assumed Growth of Seasonal Forage in a Fair Year
(pounds of dry matter per acre)

Land	Standing Crop at Beginning	Growth for Season				
		1	2	3	4	5
Unimproved range	35	65	310	5	0	20
Improvable range						
As is	90	160	780	10	0	50
Fertilize	170	280	1,420	20	0	80
Seed	140	260	1,250	15	0	75

by the following equation involving the available feed quantities:

$$SC_t + G_t = C_t + SC_{t+1}$$

where SC_t is the standing crop at the beginning of season t, G_t is the growth during season t, C_t is the forage consumed during season t, and SC_{t+1} is the standing crop at the beginning of the next season.

Every activity producing feed is potentially structured as the above difference equation requires. Hence, the activities of the static model of section 18.1 such as unimproved range (x_3); improvable range, graze as is (x_4); improvable range, fertilize (x_5); and improvable range, seed and fertilize (x_6) are specified according to the multistage recursion.

Table 18.8 illustrates the structure of the new initial tableau corresponding to the improvable range, graze as is activity. The unimproved range activity is also similarly specified. In order to associate the constraints with the corresponding variables in the given season, some of the right-hand-side coefficients are displayed with a pre- and a post-subscript. For example, $_4b_1$ corresponds to variable x_4 in growth season 1. The specification of all the other activities remains as in table 18.3. The modification of the ranch management problem from a static model to a multistage specification has produced the desired result of utilizing all the available feed produced on the ranch in every season. This objective is associated with an optimal value of the plan equal to \$41,127.89, a further substantial increase over the last attempt.

The level of positive activities constituting the optimal ranch management plan is as follows:

$$x_1 = 75.00 \quad \text{irrigated pasture}$$
$$x_3g_1 = 3,000.00 \quad \text{unimproved range, growth in season 1}$$

x_3g_2 = 3,209.26 unimproved range, growth in season 2
x_3g_3 = 281,906.69 unimproved range, growth in season 3
x_3g_5 = 36,168.01 unimproved range, growth in season 5

Table 18.8. Partial Initial Tableau for a Multistage Model of the Cattle Ranch Management Problem

Constraint	x_4g_1	x_4g_2	x_4g_3	x_4g_4	x_4g_5	x_4sc_2	x_4sc_3	x_4sc_4	x_4sc_5
b_3	1								
b_6	0.5								
. . . .
$_4b_1$	250					−1			
$_4b_2$		780				1	−1		
$_4b_3$			10				1	−1	
$_4b_4$				0				1	−1
$_4b_5$					50				1
. . . .
b_{41}									
b_{42}									
b_{43}									
b_{44}									
b_{45}									

Table 18.8. (Continued)

Constraint	x_4c_1	x_4c_2	x_4c_3	x_4c_4	x_4c_5	x_{24}	x_{25}	x_{26}
b_3								
b_6								
. . . .
$_4b_1$	−1							
$_4b_2$		−1						
$_4b_3$			−1					
$_4b_4$				−1				
$_4b_5$					−1			
. . . .
b_{41}	1					−2334	−1305	−1305
b_{42}		1				−2675	−1602	−1602
b_{43}			1			−1967	−821	−304
b_{44}				1		−948	−550	
b_{45}					1	−1470	−671	

$x_3 s c_4$	=	525,078.94	unimproved range, standing crop, season 4
$x_3 c_1$	=	300,000.00	unimproved range, consumption, season 1
$x_3 c_2$	=	994,870.62	unimproved range, consumption, season 2
$x_3 c_3$	=	884,454.56	unimproved range, consumption, season 3
$x_3 c_4$	=	525,078.94	unimproved range, consumption, season 4
$x_3 c_5$	=	723,360.31	unimproved range, consumption, season 5
$x_4 g_1$	=	1,000.00	improvable range, as is, growth in season 1
$x_4 g_2$	=	780,000.00	improvable range, as is, growth in season 2
$x_4 g_3$	=	10,000.00	improvable range, as is, growth in season 3
$x_4 g_5$	=	50,000.00	improvable range, as is, growth in season 5
$x_4 s c_4$	=	10,000.00	improvable range, as is, standing crop 4
$x_4 s c_5$	=	10,000.00	improvable range, as is, standing crop 5
$x_4 c_1$	=	250,000.00	improvable range, as is, consumption 1
$x_4 c_2$	=	780,000.00	improvable range, as is, consumption 2
$x_4 c_5$	=	60,000.00	improvable range, as is, consumption 5
x_8	=	1,000.00	brushland, cleared
x_9	=	25.00	alfalfa hay
x_{13}	=	150.00	irrigated pasture hay, season 1
x_{19}	=	391.43	purchased oat hay, season 1
x_{22}	=	200.00	sold alfalfa hay
x_{24}	=	600.00	cows
x_{25}	=	166.00	native steers

All the land is used for productive purposes, and all the forage produced on the land is either allocated to cattle or sold. Now, both cow-calf units as well as native steers are raised on the ranch with the necessity of supplementing their diet in season 1 by means of purchased oat hay. Overall, this plan seems more plausible and satisfactory than any previous one.

The dual variables corresponding to this optimal plan are:

y_1	=	41.97	irrigated pasture
y_2	=	3.68	unimproved range
y_3	=	10.00	improvable range, as is
y_4	=	1.46	brushland
y_5	=	216.17	alfalfa hay
y_6	=	0.062	capital
y_7	=	0.042	pasture hay, dry matter
y_8	=	0.044	alfalfa hay, dry matter
y_{40}	=	0.042	forage, dry matter, season 1
y_{45}	=	11.63	cow-calf

The shadow prices of the various types of land reflect their relative productivity on a per acre basis. On a pound of dry matter basis, pasture hay is valued slightly below alfalfa hay. Capital used in the ranch enterprise earns 6.2 percent of return. The marginal productivity of a cow-calf unit is $11.63 over and above the $110 dollars received on the market.

Chapter 19

The Measurement of Technical and Economic Efficiency

19.1 The Notion of Relative Efficiency

The problem of deciding the ranking of technical and economic abilities among a group of entrepreneurs is an old one and has received considerable attention from economists at different levels of sophistication. But why is it important to determine who is the best entrepreneur, who is the second best, and who is the worst? For a long time economists have assumed that the principal goal of rational entrepreneurs is maximizing profit subject to constraints imposed by the local environment. Studying the performance of various entrepreneurs, therefore, is a step toward understanding the process of making technical and economic decisions. When the best decision makers have been identified, a further presumption exists that their strategies can be used to advise those entrepreneurs who are lagging behind in the quest for maximum profit.

If it were possible to know in absolute the highest levels of technical and economic results achievable at any given time and place, the efficiency problem would not exist. No absolute standard, however, can be defined. The only feasible alternative, therefore, is to measure the technical and economic efficiency relative to a group of entrepreneurs. Not only will the ranking of performances depend on the size and characteristics of the sample, but the efficient frontier will depend upon the measuring procedure.

Technical and economic efficiencies have often been defined as follows:

technical efficiency is the ability of entrepreneurs to produce as much output as possible with a specified endowment of inputs, given the environmental conditions that surround them; economic efficiency is the ability to choose those quantities of inputs that maximize the net revenue function, given the market conditions of factor supply and product demand. The necessity of observing more than one entrepreneur, as explained above, requires the introduction of a further qualification of the two efficiency notions. Even when the sample of firms is judged to be homogeneous, in fact, the environmental and technical conditions under which entrepreneurs must operate cannot be taken to be identical for all of them. As a consequence, it simply is not fair to state that a manager in favorable environmental conditions is more efficient than others merely on the basis of a net returns measure. A proper evaluation of managerial efficiency, then, also has to account for what two managers would have achieved operating in each other's environment. The idea of *revealed efficiency* is thus born.

To implement the concept of revealed efficiency, it is convenient to distinguish two groups of factors responsible for the difference of performance among firms. They are represented by factors connected with the entrepreneur's ability to choose input quantities that maximize the net revenue function (determining the entrepreneur's behavior vis-a-vis market conditions) and factors associated with technical knowledge, the presence of specialized inputs, and the environmental conditions of the production process. These second factors are incorporated into the individual firm's production function. Two systems of analytical relations, therefore, constitute the theory's skeleton: the production function and the decision equations. Any classification of efficiency must be based upon this structure.

When empirical data are used to assess the rational behavior of economic agents, it is first necessary to consider ways to detect violations of the assumptions built into the model of such a behavior. In fact, only when using consistent data is it legitimate and possible to produce meaningful statements of economic inference. One of the main assumptions of firm theory deals with a maximizing (or minimizing) scheme. It is important to ask, therefore, whether all the producers considered in the sample actually behaved according to such an assumption. If not, hardly anything can be said about efficiency measures applied to them. Above all, it would be especially incorrect to use data from producers that failed to satisfy these assumptions as a basis for comparison with other entrepreneurs in the sample.

If we attempt to measure the managerial efficiency of two entrepreneurs, it would be ideal to test their ability when working in each other's environment before establishing their ranking. Once the information has been collected, we must rely on inferential procedures in order to unravel the story. We may first ask why the particular entrepreneur chose some specific product mix, a certain geographical location for business, a given firm's size, and a particular technology. When considered *ex ante,* a wide range of choices was available. From the actual selection we may conclude that the present results represent the consequence of the entrepreneur's best judgement exercised in order to achieve the maximizing (minimizing) objective. We would expect that in other production conditions this entrepreneur would not perform as well.

Consider a group of n firms. The efficiency relation between any two of these firms is defined as follows: firm j_1 is revealed as more efficient than firm j_2, written as

$$j_1 \boxed{\geq} j_2$$

if and only if

(A) $$\pi_{j_1}(e_{j_1}) \geq \pi_{j_1}(e_{j_2})$$

and

(B) $$\pi_{j_1}(e_{j_1}) \geq \pi_{j_2}(e_{j_1})$$

where π is a measure of firm's performance (usually profit) characterized by environmental and technical factors incorporated in the production function and e is the vector of factors determining the economic behavior of the entrepreneur.

Relation A states that the profit of firm j_1 must be higher than the profit obtainable by adopting the economic decisions of entrepreneur j_2. Relation B states that the same profit must also be higher than that obtainable using the decisions of entrepreneur j_1 in the environment of firm j_2.

For the complete ranking of firms in a given sample, the efficiency relation stated above must be extended to insure that all binary comparisons of entrepreneurs do not conflict. In other words, the efficiency relation should be antisymmetric for any pair of firms whether their efficiency was measured in a direct or indirect comparison. This condition can be expressed as follows: given n firms, if firm 1 is revealed as more efficient than firm 2, firm 2 is revealed as more efficient than firm 3, etc., firm $n-1$ is revealed as more efficient than firm n, then, firm n cannot be revealed as more efficient than firm 1. In symbolic notation, for all $j, j = 1, \ldots, n,$

(C) $\quad j_1 \boxed{\geq} j_2 \boxed{\geq} j_3 \ldots j_{n-1} \boxed{\geq} j_n,$ implies $j_n \boxed{\not\geq} j_1$

A meaningful measure of efficiency is possible only for firms whose technical and economic information does not violate the above relations. The implementation of this relative efficiency concept requires the following data: (a) quantities of inputs and outputs chosen by each entrepreneur and the corresponding prices prevailing in that environment and (b) the individual firm technology. Although the collection of both types of information is possible, it is customary to limit firm surveys to information of the first type. In this case, the measurement of efficiency requires further assumptions. Typically, all the firms in the given sample are assumed to operate under the same technology.

19.2 The Farrell Method

In a pioneering paper that appeared in 1957, Michael Farrell proposed to measure the efficiency of a sample of firms through a set of assumptions that could be implemented by linear programming techniques. Farrell specified three types of efficiency: technical, economic, and overall. To illustrate the main ideas underlying the method, let us consider two inputs and one output. Given a sample of firms, each characterized by the same production function of output Y, which is assumed to be linearly homogeneous with respect to capital C and labor L, it is possible to define the *unit-output isoquant* as

$$1 = F\left[\frac{C}{Y}, \frac{L}{Y}\right]$$

The unit-output isoquant is, therefore, the set of input-output combinations characterized by the highest level of technical efficiency. In figure 19.1, the unit-output isoquant is represented by the $I - I'$ curve. All the firms with capital/output and labor/output ratios strictly above the $I-I'$ curve, at the point $P(c_p, l_p)$, are technically inefficient. The index of technical efficiency (TE) is defined at the ratio between the distance from the origin of the unit-output isoquant and the distance from the origin of the given firm's normalized input combination. With reference to figure 19.1, the index of technical efficiency is defined as

$$\text{technical efficiency} \equiv TE \equiv \frac{OM}{OP}$$

Economic efficiency (sometimes called price efficiency) is defined with reference to the unit-output isocost represented by the $A - A'$ line in figure 19.1. A point on that line, such as G, represents an input-output combination that is efficient from an economic viewpoint but not from a technical one. The point Q is technically and economically efficient. The index of economic efficiency (EE) is thus defined as

$$\text{economic efficiency} \equiv EE \equiv \frac{OG}{OM}$$

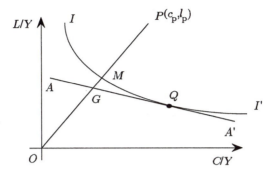

Figure 19.1. Technical and economic efficiency.

Overall efficiency (OE) is defined as the combination of technical and economic efficiencies, that is,

$$\text{overall efficiency} \equiv OE \equiv TE \times EE = \frac{OM}{OP} \times \frac{OG}{OM} = \frac{OG}{OP}$$

The range of all three indexes is between 0 and 1. From an operational standpoint, the main problem is constituted by the measurement of the unit-output isoquant. Farrell suggested a general outline of a computational technique without specifying the details and without presenting an efficient computer program. This task was left for Jim Boles.

19.3   Linear Programming and Efficiency Indexes

Let us consider a group of n firms whose efficiency indexes must be computed. In the Farrell procedure, the input quantities of each firm, normalized by the corresponding output, form the technical coefficients in a series of n linear programming models. The coefficients of the objective functions are the unit-output quantities produced by each firm. The constraints of the n LP problems are given by the normalized input quantities of each firm member of the group. Thus, to compute the efficiency indexes of n firms, we must solve n distinct linear programming problems.

To illustrate the procedure in greater detail, let us consider three firms that produce the same output, x, by means of two inputs, b_1 and b_2. The matrix A of the three linear programming problems is constructed as

$$\begin{bmatrix} \dfrac{b_{11}}{x_1} & \dfrac{b_{12}}{x_2} & \dfrac{b_{13}}{x_3} \\[2mm] \dfrac{b_{21}}{x_1} & \dfrac{b_{22}}{x_2} & \dfrac{b_{23}}{x_3} \end{bmatrix} \quad \text{or} \quad \begin{bmatrix} a_{11} & a_{12} & a_{13} \\[2mm] a_{21} & a_{22} & a_{23} \end{bmatrix}$$

where the second index identifies the firm and $a_{ij} = b_{ij}/x_j$, $i = 1$, 2, and $j = 1$, 2, 3. Then the jth linear programming problem is specified as follows:

Primal maximize $V_{0j} = \quad v_1 + \quad v_2 + \quad v_3$

$$\text{subject to} \quad a_{11}v_1 + a_{12}v_2 + a_{13}v_3 \leq a_{1j}$$
$$a_{21}v_1 + a_{22}v_2 + a_{23}v_3 \leq a_{2j}$$

$$v_k \geq 0, \, k = 1, 2, 3$$

The variable v_k represents the output quantity producible by the kth firm given the input availabilities of the jth firm. The rationale of the Farrell method exploits the fact that the input quantities of the jth firm produce, by construction, a unit-output isoquant. Suppose $j = 1$. Then, if the production process used by firm #1 is technically efficient, the solution of the above linear programming problem will be $v_1 = 1$, $v_2 = v_3 = 0$. Hence, the maximum value of the objective function is $V_{01} = 1$. On the contrary, if both firms #2 and #3 are technically more efficient than firm #1, the LP solution will be $v_1 = 0$, and $v_2 + v_3 > 1$, with $V_{01} > 1$. This second instance corresponds to the case where, combining the technical processes of firms #2 and #3 with the input availabilities of firm #1, it is possible to produce a level of output greater than the unit quantity. A convenient index of technical efficiency for the jth firm, then, is defined as

$$\text{technical efficiency} \equiv TE_j = \frac{1}{V_{0j}}$$

Figure 19.2 illustrates the unit-output isoquant constructed from the linear programming interpretation of the Farrell method. In a 2-dimensional space there are as many efficiency cones as there are efficient firms minus one. A cone, then, corresponds to an efficient LP basis for the firms falling within it. Firm #1, for example, is technically less efficient than both firms #2 and #3. For this reason, its representation point P_1 lies above the P_2P_3 segment in the cone generated by activities $\mathbf{a_2}$ and $\mathbf{a_3}$. Firm #1 (and all the firms falling in the same cone) must be compared with the P_2P_3 segment of the unit-output isoquant for the purpose of measuring its technical efficiency.

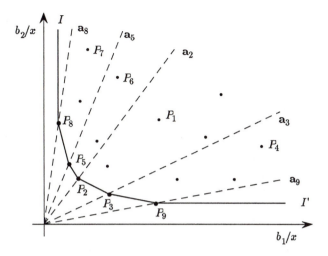

Figure 19.2. Technical efficiency from a LP interpretation of the Farrell method.

The same index of technical efficiency can be obtained from the dual problem:

Dual minimize $W_{0j} = a_{1j}w_1 + a_{2j}w_2$

subject to

$$a_{11}w_1 + a_{21}w_2 \geq 1$$
$$a_{12}w_1 + a_{22}w_2 \geq 1$$
$$a_{13}w_1 + a_{23}w_2 \geq 1$$

$$w_1, \ w_2 \geq 0$$

In the dual formulation, technical efficiency is achieved when resources of the jth firm are combined in minimum quantities (through the scaling factors w_1 and w_2) to achieve the unit-output isoquant subject to the condition that the other firms in the group also be classified on or above the unit level of production. When this objective is reached, $W_{0j} = 1$, and the jth firm is classified as technically efficient. When $W_{0j} > 1$, the jth firm is technically inefficient.

To generalize the LP specification of the Farrell method, let us consider n firms each producing only one and the same commodity by means of the same m inputs, with $m < n$. The primal problem whose solution gives the index of technical efficiency of the jth firm is

Primal maximize $V_{0j} = \mathbf{u}'\mathbf{v}_j$

subject to $\qquad A\mathbf{v}_j \leq \mathbf{a}_j$

$$\mathbf{v}_j \geq \mathbf{0}$$

where $\mathbf{a}_j' = (a_{1j}, a_{2j}, \ldots, a_{mj})$ is the normalized input vector of the jth firm, $\mathbf{u}' = (1, 1, \ldots, 1)$ is the sum vector, and $\mathbf{v}_j' = (v_{1j}, v_{2j}, \ldots, v_{mj})$ is the jth solution vector. The A matrix is the same for all the n linear programs and is formed by the n vectors of normalized inputs $\mathbf{a}_k$, $k = 1, \ldots, n$. The corresponding dual problem is

Dual minimize $W_{0j} = \mathbf{a}_j'\mathbf{w}_j$

subject to $\qquad A'\mathbf{w}_j \geq \mathbf{u}$

$$\mathbf{w}_j \geq \mathbf{0}$$

where $\mathbf{w}_j$ is the vector of scaling factors associated with the jth normalized input vector. The extension of the Farrell method to cover several inputs is thus straightforward. Furthermore, the actual computation of the efficiency indexes for the n firms does not require the solution of n linear programming problems. The algorithm developed by Boles takes advantage of the fact that once the first linear program has been solved, the remaining $(n-1)$ indexes are computed using postoptimization techniques of the type discussed in chapter 13, since only the RHS vector needs to be changed from firm to firm.

The measurement of technical efficiency does not involve prices of any kind. Economic and overall efficiency, on the contrary, require the additional information about market output and input prices received and paid by entrepreneurs. Given a competitive market environment, a unique set of prices is assumed to prevail for all the n firms. Then, overall efficiency corresponds to a technically efficient combination of resources that minimizes the cost of the unit value of production, as seen by point Q in figure 19.1. Let C be this unit cost defined as

$$C = \frac{\text{total input cost}}{\text{value of production}}$$

Then the definition of overall and economic efficiencies is as follows:

$$\text{overall efficiency} \equiv \quad OE_j = \frac{\min_k C_k}{C_j}$$

$$\text{economic efficiency} \equiv \quad EE_j = \frac{OE_j}{TE_j} = \frac{\min_k C_k}{TE_j C_j}$$

Therefore, $OE_j = TE_j \times EE_j$, as required by the Farrell method. These definitions of overall and economic efficiency involving the average cost rather than the marginal cost are legitimate in this case because the LP technology is linear, as discussed in section 1.5. Although it is possible to find that several firms are technically efficient, overall efficiency will, in general, be associated with only one firm as illustrated by point Q in figure 19.1. The computation of the indexes of economic and overall efficiency does not require the use of linear programming, since the selection of the minimum unit cost is easily done by inspection.

19.4 Technical and Economic Efficiency in a Group of Small Family Farms

The Farrell method was applied to a group of small family farms located in the Campania region of Italy. These farms have been the subject of a long-term study carried out by the Advanced Training Center for Agricultural Economics of the University of Naples with the purpose of identifying management strategies for technical assistance to the local farmers. The group of 14 farms selected for the efficiency study is rather homogeneous with respect to size, agricultural practices, and economic environment. They produce a variety of outputs ranging from animal products (milk, veal, hogs, poultry) to crops such as wheat, tomatoes, sugar beets, tobacco, and alfalfa. A single index of total production was defined by aggregating all outputs with average prices. Four input categories (land, labor, dairy cows, and miscellaneous inputs) were also defined. Miscellaneous inputs included the service flow of machinery and equipment, feedstuff, fertilizers, pesticides, and hired services. The information suitable for computing the efficiency indexes is displayed in table 19.1. The four input categories are expressed in terms of the corresponding aggregate output (defined as the value of saleable production) as required by the Farrell method.

The indexes of technical efficiency were computed following the primal version of the Farrell method, as described in the previous section. The indexes of economic and overall efficiency were computed using average market input prices: 90,000 Lire/hectare for land, 1,050,000 Lire/man-year for labor, and 110,000 Lire/cow for dairy cows.

The three series of indexes are given in table 19.2. Each farm is also associated with reference farms (whose activities are part of the optimal basis) to which it was compared on the unit-output isoquant by the linear programming algorithm. Hence, an important piece of information obtained from the application of this approach is an indication of the farms to be used as models for improving the technical efficiency of the jth farm. Consider, for example, farm #4. With an index of technical efficiency of 0.781,

Table 19.1. Information for the Application of the Farrell Method
(single output, normalized inputs)

Farm	Land (hectares/ output)	Labor (man-year/ output)	Dairy Cows (cows/ output)	Miscellaneous (million Lire/ output)
1	1.32	0.42	1.05	0.42524
2	1.20	0.51	1.42	0.25472
3	1.47	0.43	1.11	0.25570
4	2.28	0.71	2.22	0.22860
5	2.16	0.44	2.40	0.39805
6	2.98	0.44	1.22	0.32108
7	2.33	0.73	2.05	0.32434
8	1.86	0.48	1.85	0.16318
9	1.97	0.64	1.78	0.26726
10	1.96	0.79	1.83	0.51722
11	1.86	0.56	1.01	0.32948
12	1.86	0.54	1.51	0.26748
13	2.04	0.90	0.88	0.21781
14	2.19	0.49	1.75	0.57192

its direct models are farms #2, #8, and #13. Analogous guidelines are established for any other farm. It is of interest, indeed, to notice that farms associated with a maximum level of technical efficiency also have their own models for improvement. For example, the reference farms for farm #3, which shows an index of technical efficiency equal to 1.0, are farms #1 and #2 (besides, of course, itself). This fact suggests that there are multiple solutions to the problem of technical efficiency.

A further item of information pertaining to the relative scarcity of the four inputs as they have been selected in the various farms is given by the shadow prices presented in table 19.3. The information can be classified in two parts: with respect to efficient farms (technical efficiency of 1.000) and relative to the inefficient ones. Efficient farms (1, 2, 3, 8, 11, and 13) have the exact relative mix of inputs. Some of them, however, are associated with a dual degeneracy suggesting the possibility of different valuations of the existing resources. Inefficient farms do not exhibit any dual degeneracy. Some of them (4, 5, 6, 10, and 14) indicate the presence of surplus inputs suggesting that one way of improving the technical efficiency is to readjust the input combination. The remaining inefficient farms (7, 9, and 12) do not exhibit any special imbalance of the input combination (all the shadow prices are greater than zero), indicating that the improvement in technical

Table 19.2. Efficiency Indexes for a Group of Small Family Farms
(single output)

Farm	Reference Farms	Technical Efficiency	Economic Efficiency	Overall Efficiency	Cost per Lira of Saleable Production
1	1, 2	1.000	0.868	0.868	1.095
2	2	1.000	0.913	0.913	1.042
3	1, 2, 3	1.000	1.000	1.000	0.951
4	2, 8, 13	0.781	0.863	0.674	1.412
5	1, 3	0.960	0.767	0.737	1.290
6	1, 3	0.970	0.837	0.811	1.172
7	2, 3, 8, 13	0.683	0.705	0.632	1.505
8	8, 13	1.000	0.937	0.937	1.015
9	2, 3, 8, 13	0.812	0.903	0.733	1.298
10	1, 2, 3	0.657	0.845	0.555	1.713
11	3, 11, 13	1.000	0.798	0.798	1.191
12	2, 3, 8, 13	0.859	0.955	0.819	1.161
13	1, 13	1.000	0.666	0.666	1.428
14	1	0.857	0.763	0.654	1.455

efficiency must be found in a proportional reduction of the input mix.

To study more directly the success of the efficient farms and possible improvement strategies for the inefficient ones, it is convenient to refer any indication from the analysis of technical efficiency to table 19.4, which presents the levels of inputs and outputs in their natural units. The asterisk indicates technically efficient firms. By inspecting this table it is clear that neither the absolute level of the various input categories (whether high or low) nor the level of the saleable production is a sufficient indication of successful performance. Farm #1, in fact, has a high level of miscellaneous expenditures while farm #3 has a low one, but both are technically efficient. Conversely, farm #14 and #4 have a high and low level of miscellaneous inputs, respectively, and both are technically inefficient. The difficulty of sorting out a pattern of determinants of technical efficiency justifies the need for a more comprehensive approach like the Farrell method.

The indexes of economic and overall efficiency are computed according to the formulas involving the ratio of the minimum unit cost of the efficient farm relative to that of the jth farm. In this sample, farm #3 is the cost-efficient unit (see the fifth column of table 19.2). It is interesting to notice that some farmers are more capable from a technical point of view while others show more ability to deal with their economic environment.

Table 19.3. Shadow Prices of Technical Efficiency

Farm	Land 1	Labor 2	Dairy Cows 3	Miscellaneous 4	Dual Degeneracy
1	0.602	0.000	0.195	0.000	2, 4
2	0.833	0.000	0.000	0.000	3, 4, 5
3	0.364	0.000	0.319	0.435	2
4	0.194	0.000	0.256	2.137	–
5	0.000	2.247	0.000	0.133	–
6	0.000	2.247	0.000	0.133	–
7	0.186	0.023	0.163	2.097	–
8	0.000	0.000	0.211	3.740	1, 2
9	0.186	0.023	0.163	2.097	–
10	0.364	0.000	0.319	0.435	–
11	0.000	0.363	0.692	0.299	1
12	0.186	0.023	0.163	2.097	–
13	0.173	0.000	0.734	0.000	2, 4
14	2.381	0.000	0.000	0.000	–

Table 19.4. Inputs and Saleable Production (natural units)

Farm	Land (hectares)	Labor (man-years)	Dairy Cows (equiva-lents)	Miscellaneous (million Lire)	Saleable Production (million Lire)
*1	6.28	2.00	5.00	2.022	4.754
*2	4.31	1.84	5.10	0.917	3.958
*3	5.16	1.50	3.90	0.897	3.400
4	6.17	1.92	6.00	0.617	2.688
5	8.65	1.75	9.60	1.595	4.237
6	12.49	1.85	5.10	1.345	4.188
7	7.96	2.50	7.00	1.106	3.404
*8	5.87	1.50	5.80	0.515	3.255
9	7.75	2.50	7.00	1.049	3.785
10	7.39	3.00	6.90	1.952	3.520
*11	7.73	2.33	4.20	1.368	4.058
12	6.89	2.00	5.60	0.991	3.834
*13	9.90	4.37	4.30	1.059	4.767
14	13.35	3.00	10.70	3.491	6.103

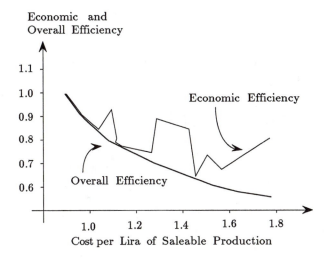

Figure 19.3. Relations among economic efficiency, overall efficiency, and cost/saleable production.

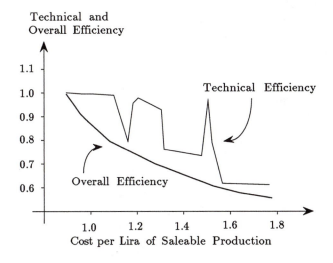

Figure 19.4. Relations among technical efficiency, overall efficiency, and cost/saleable production.

The relationships among economic, technical, and overall efficiency and the unit cost of production are displayed in figures 19.3 and 19.4. The overall efficiency curve is, of course, the envelope of the technical and economic efficiencies. The two figures make abundantly clear the necessity of achieving high levels of both technical and economic efficiency in order to fulfill the goal of maximum profits.

19.5 Efficiency in Multiproduct Firms

In previous sections, the firms were assumed to produce a single output. When analyzing the 14 Italian farms, therefore, grouping of all products into a single index of production was necessary. This extreme aggregation requires stringent conditions that could not be verified on this sample of information. When dealing with agricultural firms, therefore, it is of interest to evaluate their efficiency by keeping the production at a satisfactory level of disaggregation.

Let us suppose, then, that the n firms to be classified according to their technical efficiency produce p products by means of m inputs. The use of linear programming allows for an easy extension of the Farrell method to this more complex case. In the case of more than one product the normalization is optional: the same efficiency indexes are obtained with either normalized or unscaled variables. This invariance of the efficiency measures is a valuable property of the Farrell method. It is convenient, therefore, to develop a procedure without the normalization of the output and input quantities as performed in section 19.3.

Let $\mathbf{x}_j$ and $\mathbf{b}_j$ represent the vectors of output and the input quantities of the jth firm, and let θ_j be a nonnegative scalar. The linear programming model that produces an index of technical efficiency for the jth multiproduct firm is specified as follows:

Primal maximize θ_j

$$\text{subject to} \quad (1 + \theta_j)\mathbf{x}_j \; - \; X\mathbf{v}_j \; \leq \; \mathbf{0} \quad \text{output constraints}$$

$$B\mathbf{v}_j \; \leq \; \mathbf{b}_j \quad \text{input constraints}$$

$$\theta_j \geq 0, \; \mathbf{v}_j \geq \mathbf{0}$$

where X and B are the matrices of outputs and inputs, respectively, for all the n firms. The index of technical efficiency for the jth multiproduct firm is now defined as

$$\text{technical efficiency} \equiv TEMP_j \; = \; \frac{1}{(1 + \theta_j^*)}$$

where θ_j^* is the optimal value of the objective function in the above linear program. If $\theta_j^* = 0$, the jth firm is technically efficient; if $\theta_j^* > 0$, it is not.

For a convincing interpretation of the above specification it is useful to discuss the limiting case when the n firms produce a single output and the vector $\mathbf{x}_j$ becomes a scalar x_j. This is the scenario already presented in section 19.3 and, therefore, the two specifications must give identical results. The formal difference between the two LP problems is represented by the objective function of the normalized model in section 19.3, which has become the first constraint in this (not normalized) section. If the jth firm is technically efficient according to the normalized model, $V_{0j} = \mathbf{u}'\mathbf{v}_j = 1$ and, since $\mathbf{u}'\mathbf{v}_j = 1$ is the representation of a unit product of an efficient firm in the normalized system, it corresponds to $x_j = \mathbf{x}'\mathbf{v}_j$ in the system without normalization. This relation is, in fact, given by the first constraint of the primal in this section, with $\theta_j = 0$. When $V_{0j} > 1$ in the normalized LP model, the significance of this event is that a linear combination of technical processes belonging to other firms exists in the sample, which could produce a level of output greater than unity when using the input quantities of the jth firm. Thus, this firm is technically inefficient. In the model without normalization, the same event means that it is possible to find $\theta_j^* > 0$ such that $(1 + \theta_j^*)x_j = \mathbf{x}'\mathbf{v}_j$, the first constraint of the LP system.

The discussion of the dual problem is also illuminating. In order to specify the dual problem correctly, it is convenient to break up the $(1 + \theta_j)$ term in the first constraint of the primal and to write the vector $\mathbf{x}_j$ explicitly on the right-hand side also.

Dual minimize $W_{0j} = -\mathbf{x}_j'\mathbf{w}_j + \mathbf{b}_j'\mathbf{y}_j$

$$\text{subject to} \quad \mathbf{x}_j'\mathbf{w}_j \geq 1$$

$$-X'\mathbf{w}_j + B'\mathbf{y}_j \geq 0$$

$$\mathbf{w}_j \geq 0,\ \mathbf{y}_j \geq 0$$

where $\mathbf{w}_j$ and $\mathbf{y}_j$ are the vectors of shadow weights of the outputs and inputs of the jth firm, respectively.

The interpretation of the dual objective function corresponds to the minimization of the difference between the input combination, $\mathbf{b}_j'\mathbf{y}_j$, and the product mix, $\mathbf{x}_j'\mathbf{w}_j$, that the jth firm could have realized on the basis of a comparison with the best firms in the group. The first dual constraint states that the jth firm's product mix must be greater than or equal to one. The second set of dual constraints specifies the equilibrium conditions: for each firm, the input combination, $B'\mathbf{y}_j$, must be greater than or equal to the product mix, $X'\mathbf{w}_j$. Obviously, each entrepreneur will attempt to minimize loss, even when it is measured in physical terms. For the entrepreneur of the jth firm, however, there exists a further constraint given by the comparison of ability to produce a weighted unit of output with that of the most efficient entrepreneurs in the group. If the entrepreneur can be classified as efficient, the value of the dual objective function will be $W_{0j} = 0$, otherwise the solution of the dual problem will show that the total possible input combination is greater than the possible product mix and, therefore, $W_{0j} > 0$.

The indexes of overall efficiency for the multiproduct case are the same as those for a single product. A possible exception is when information about the allocation of inputs to the different products is available. In that case, indexes of overall efficiency can be computed in relation to each product. The index of economic efficiency depends upon the index of technical efficiency as well as that of overall efficiency.

19.6   An Example of Efficiency in Multiproduct Farms

The same Italian farms were tested for technical and economic efficiency under a more disaggregated output classification. In particular, three categories of products were defined: animal products, which included milk and veal/beef meat; crop products including wheat, tomatoes, sugar beets, tobacco, etc.; and other products including hogs, poultry, rabbits, etc. The detailed information is presented in table 19.5. These data, together with the input information of table 19.4, are used to implement the proper linear programming specification for this multiproduct efficiency analysis whose results are reported in table 19.6.

In comparison to table 19.2, several more farms are now classified as technically efficient. This fact is due primarily to the expansion from a unit-output curve associated with table 19.2 to a three-dimensional surface now representing the unit-output isoquant. With more dimensions, it is natural for the Farrell method to employ more farms in the definition of the unit-output isosurface. These results are also an indication that, indeed, the 14 family farms should be classified as multiproduct farms. It is incontestable, however, that farms #7, #9, and #10 are technically inefficient.

Only farm #3 exhibits overall efficiency. This is consistent with economic analysis as illustrated by figure 19.1. Finally, a study of the reference farms shows that even efficient units can adopt the technical processes of other basic farms. It means that the problem of selecting a technically efficient combination of inputs has multiple solutions. This aspect of the Farrell method is of great empirical relevance because we often observe that real-life problems have more than one solution.

As previously mentioned, the measure of technical efficiency in the

Table 19.5. Product Information for the Farrell Method

Farm	Animal Products (million Lire)	Crop Products (million Lire)	Other Products (million Lire)
1	1.954	2.335	0.465
2	2.284	0.708	0.606
3	2.102	0.860	0.438
4	2.043	0.120	0.525
5	3.400	0.213	0.624
6	2.152	1.315	0.721
7	2.762	0.288	0.354
8	2.608	0.288	0.359
9	2.902	0.623	0.260
10	2.641	0.480	0.399
11	1.812	1.583	0.663
12	2.773	0.304	0.757
13	1.858	2.282	0.627
14	4.236	0.561	1.306

presence of multiple products is invariant to the choice of the normalizing output. For this reason, we can avoid normalization altogether, as in table 19.6. It may be of interest to know, however, that the computation of efficiency indexes under normalization by one of the three product categories changes the reference farms. This is explained by the fact that, in this case, the LP basis has one dimension less than that without normalization.

19.7   Efficiency and Returns to Scale

The discussion in previous sections was based upon the hypothesis that all the firms in the group exhibit constant returns to scale. This assumption is often restrictive and may bias the efficiency measurements. The alternative is to further classify the n firms into homogeneous subgroups with respect to the scale of operation. When discussing the scale of firms, economists distinguish two different notions of it: *economies of scale* and *returns to scale*. Economies of scale are, in turn, classified as either real or pecuniary. As one can easily see, economies of scale is a notion that is rather difficult to define precisely and even more problematic to measure. We will not attempt to either define it or measure it. Returns to scale, on the contrary, is a technical property of technologies. It refers to the relationship between

Table 19.6. Efficiency Indexes for a Sample of Multiproduct Family Farms
(without product normalization)

Farm	Reference Farms	Technical Efficiency	Economic Efficiency	Overall Efficiency
1	1	1.000	0.868	0.868
2	2, 3, 5, 6, 12, 14	1.000	0.913	0.913
3	1, 3, 13	1.000	1.000	1.000
4	4, 8, 12	1.000	0.674	0.674
5	2, 3, 5, 12, 14	1.000	0.737	0.737
6	2, 6, 11, 12	1.000	0.811	0.811
7	2, 3, 8	0.807	0.783	0.632
8	2, 8	1.000	0.937	0.937
9	2, 3, 8	0.860	0.852	0.733
10	2, 3	0.780	0.711	0.555
11	1, 3, 11, 13	1.000	0.798	0.798
12	11, 12, 13	1.000	0.819	0.819
13	1, 2, 11, 13	1.000	0.666	0.666
14	5, 6, 12, 14	1.000	0.654	0.654

inputs and output as all inputs are varied in the same proportion. There are three types of returns to scale: constant, increasing, and decreasing. Consider the production function $q = f(x_1, x_2, \ldots, x_n)$. When multiplying all the inputs by the same factor, t, one of the following three cases of returns to scale occurs:

$$f(tx_1, tx_2, \ldots, tx_n) \; = \; tf(x_1, x_2, \ldots, x_n) = tq \qquad \text{constant returns}$$

$$f(tx_1, tx_2, \ldots, tx_n) \; < \; tf(x_1, x_2, \ldots, x_n) = tq \qquad \text{decreasing returns}$$

$$f(tx_1, tx_2, \ldots, tx_n) \; > \; tf(x_1, x_2, \ldots, x_n) = tq \qquad \text{increasing returns}$$

Linear programming assumes that constant returns to scale prevail. In a sample of farms, however, this assumption may be violated for many different reasons, no matter how careful the selection has been. It is of interest, therefore, to classify the sample according to an index of returns to scale that reflects the empirical information available. This procedure can be specified as follows: assuming the firms analyzed produce a single output, the linear programming model that verifies the returns to scale is

Primal maximize $V_{0j} = \mathbf{u}'\mathbf{v}_j$

$$\text{subject to} \quad A\mathbf{v}_j \leq \mathbf{a}_j$$

$$S'\mathbf{v}_j = \mathbf{0}$$

$$\mathbf{v}_j \geq \mathbf{0}$$

where A, $\mathbf{u}$, $\mathbf{v}_j$, and $\mathbf{a}_j$ are defined as in section 19.3 and where S is a matrix related to the scale of operation of the jth firm. Its definition is

$$S = \begin{bmatrix} (s_{11} - s_{1j}) & (s_{12} - s_{1j}) & \cdots & (s_{1n} - s_{1j}) \\ (s_{21} - s_{2j}) & (s_{22} - s_{2j}) & \cdots & (s_{2n} - s_{2j}) \\ \cdots & \cdots & \cdots & \cdots \\ (s_{m1} - s_{mj}) & (s_{m2} - s_{mj}) & \cdots & (s_{mn} - s_{mj}) \end{bmatrix}$$

where s_{ij} is the ith input of the jth firm measured in natural units. The set of equations specified by the constraints $S'\mathbf{v}_j = \mathbf{0}$ can be more explicitly restated for every ith input as

$$s_{ij} = \frac{\sum_{k=1}^{n} s_{ik} v_{kj}}{\sum_{k=1}^{n} v_{kj}} = \frac{\sum_{k=1}^{n} s_{ik} v_{kj}}{v_{0j}}$$

that is, the set of farms defining an efficient basic representation for the jth farm must have a weighted average input combination equal to the scale of the jth farm. All the firms with the same (or nearly the same) index of output V_0 belong to the same returns to scale subgroup, and scale-corrected efficiency indexes can be computed for the members of that group. This specification means that there will be as many efficient unit isoquants as there are different returns to scale.

The dual problem corresponds to

Dual minimize $W_{0j} = \mathbf{a}_j'\mathbf{w}_j$

$$\text{subject to} \quad A'\mathbf{w}_j + S'\mathbf{y}_j \geq \mathbf{u}$$

$$\mathbf{w}_j \geq \mathbf{0}, \ \mathbf{y}_j \text{ free}$$

The farms analyzed in previous sections are considered rather homogeneous with respect to several aspects, including the returns to scale. Nevertheless, we have applied the framework discussed here to them as an illustration of the procedure. The results are presented in table 19.7.

Table 19.7. Returns to Scale in a Group of Small Family Farms

Farm	Reference Farms	Returns to Scale	Shadow Prices of Unscaled Inputs			
			Land	Labor	Cows	Misc.
1	1, 2, 3	1.000	−0.039	0.000	0.143	0.000
2	1, 2, 3	1.000	0.128	0.000	−0.092	0.000
3	1, 2, 3	1.000	0.000	0.000	0.154	−0.060
4	2, 4, 8, 9, 13	1.000	0.197	−0.766	−0.356	1.453
5	3, 5, 6, 14	1.000	−0.005	0.327	−0.106	0.000
6	1, 6, 13, 14	1.000	−0.089	0.050	0.051	0.000
7	3, 5, 6, 9, 13, 14	1.163	0.001	−0.051	−0.113	0.215
8	2, 6, 8, 13, 14	1.000	−0.131	0.088	−0.023	0.264
9	2, 5, 8, 9, 13	1.000	−0.115	0.058	−0.038	0.276
10	2, 9, 10, 13, 14	1.000	0.047	−0.338	−0.183	0.191
11	1, 3, 6, 11, 13	1.000	−0.137	0.101	0.112	0.054
12	3, 5, 6, 8, 9, 13	1.017	0.000	−0.084	−0.125	0.275
13	1, 6, 13, 14	1.000	−0.130	0.072	0.074	0.000
14	1, 14	1.000	−0.056	0.000	0.000	0.000

As anticipated, almost all the farms show the same returns to scale, which justifies their treatment as one homogeneous group. Only farm #7 seems to operate on a slightly different scale, perhaps not sufficient for establishing a separate subgroup. For completeness, the scales of farm #7 and farm #12 can be expressed as the weighted average of the scales of the following farms:

farm #7		farm #12	
farm #3	0.026	farm #3	0.418
farm #5	0.095	farm #5	0.032
farm #6	0.018	farm #6	0.055
farm #9	0.962	farm #8	0.057
farm #13	0.057	farm #9	0.440
farm #14	0.005	farm #13	0.014

The illustration of the model indicates that it is possible to verify the returns-to-scale dimension and then to compute efficiency indexes among the farms belonging to the same scale subgroup.

The meaning of the shadow prices of the unscaled inputs is related to the relative internal tensions of the jth farm for keeping the various inputs in line with the scale dimension of the group. In general, a negative shadow price should be interpreted as a signal for reducing that input relative to the other three, while a positive shadow price suggests an increase.

Chapter 20

Decentralized Economic Planning

20.1 When Is Planning Necessary?

Entrepreneurs conduct economic activities finalized to their objectives and interest. When the objectives of different economic agents diverge, conflicts arise. Sometimes, the marketplace can mediate these conflicts through largely unconscious and seemingly haphazard efforts, especially when those interests are held with mild intensity. Often, conflicts must be mediated by an explicit agreement among the various actors who may delegate to a higher authority the task of resolving them. More often, finally, a higher authority assumes the responsibility of imposing on its subjects the resolution of conflicts. Consider a large corporation organized into various departments: procurement, production, marketing, financial, R & D, and personnel. All department managers are usually told to establish departmental goals of performance in their own interest and that of the department and are vaguely aware that they must compete with the other managers in the allocation of the corporation's resources. In this situation, conflicts inevitably arise and the board of directors (or its chair) will act as a final decision maker in the interest of the corporation as a whole.

An economy has often been likened to a giant corporation, organized in sectors that operate largely as departments, establishing sectoral goals in isolation from other sectors and competing with them for the allocation of national resources such as energy, capital, and strategic raw materials. To sectoral objectives there corresponds a "common good" interest at the economywide level. Who the macroagent, or agency, that defines the common good objective for the economy should be (the invisible hand or the

307

central planning board?) has been the subject of fierce debates for at least one hundred years. Often, the debate has become ideological and, even worse, proponents as well as opponents of economic planning have found excuses for imposing their opinions through bloody coercion.

Throughout the decades, therefore, economic planning has been heralded as a panacea for solving the problems of unbalanced economies, feared as a scheme for depriving individuals of their unrestricted freedom, glorified as the highest level of societal organization, and vituperated as a manmade misfortune. A quiet reflection on the literary and political vicissitudes of economic planning suggests that its characteristics and working have, more often than not, been misunderstood and thus badly implemented.

It seems obvious that the marketplace is incapable of mediating in a satisfactory way all the conflicts arising from the decisions of unbridled economic agents. It is clear, on the other hand, that economic planning may often act as the classical bull in the china shop. It is unfortunate that economic theory seems reduced to debate which is the lesser evil but, perhaps, this is another dimension of economics as a dismal science.

20.2 Decentralized Economic Planning and LP

This chapter will focus upon the inherent, but not inevitable, misunderstanding associated with economic planning. We will use linear programming to clarify a crucial aspect of the planning process that, very likely, has contributed to the degeneration observed in several instances. This use of LP may be surprising, at first, but its dual structure is eminently suited for exposing, in a purely logical way, the fragile nature of planning in general and economic planning in particular.

Economic planning is, first of all, a gathering of information. After the plan has been formulated, it will be implemented. We are concerned here only with the formulation of a stylized economic plan. The information exchanged between sources can be either true or false. False information is usually called misinformation and it introduces into the planning process a further psychological aspect (detecting and punishing liars) that we disregard entirely. We assume, therefore, that the information collected by the various parties involved is true and complete.

When it comes to economic planning, almost everybody (from politicians to news people, from corporation executives to workers' unions), feels entitled to express an expert opinion. This is an indication of how central to peoples' life and how contested the subject is. A disclaimer is thus in order. This chapter does not entitle the reader (and the writer) to any feeling of expertise about economic planning; it contains only an initial understanding of the difficulties of assembling a plan and maintaining a general consensus around it, without which any successful implementation is doomed from the start.

Decentralized economic planning is a process that divides the responsibility of collecting the information needed for formulating the plan between a center (central planning board) and the periphery (sectors). We will not discuss centralized planning, which assumes, by definition, that all the information already resides at the center. Centralized planning is utopian and, therefore, uninteresting.

Collection of information implies an exchange of information. The typical way to gather information, in fact, is to stimulate a reaction of the subject via the proposal (or the imposition) of a message in the form of an experiment. Hence, the central planning board (CPB) has the option of either informing the sectors about the prices of certain strategic resources or declaring that each sector should produce preassigned quotas of outputs requiring preassigned quotas of strategic resources. These two alternatives constitute the fundamental schemes of decentralized planning. As the reader can surmise, they are dual to each other in the sense of this book, where quantities and prices have been emphasized to be dual entities. It is customary to associate with western economies a type of economic planning where the CPB uses the prices of strategic resources to elicit the required information from the sectors. We will call this planning process the *primal scheme.* Alternatively, a type of economic planning that relies upon the issuance of commodity quotas (strategic inputs and outputs) by the CPB in order to elicit the needed information is referred to as the Russian scheme, simply because the Soviet Union was the first to explicitly adopt this process on a massive scale. We will call this planning process the *dual scheme.* The labeling, of course, is only a means to quickly evoke the description of the process. In reality, both western and Russian planners have used a combination of prices and quotas to obtain the necessary information. But we are not discussing the reality of economic planning, only its logical structure.

The important conclusion of this analysis should be anticipated at this point. Regardless of the planning scheme used for gathering the information, the final stage of the plan corresponds to the issuance of both prices and quotas by the CPB to the sectors. This is the stage where the symmetry of the planning process breaks down and the consensus between CPB and sector managers may disappear. This asymmetry constitutes the fragile aspect of economic planning detected at a purely logical level.

20.3   The Western Scheme of Economic Planning

Whether at the firm or the economywide level, economic activity requires knowledge of the feasible set of production. In the western scheme of economic planning, the CPB starts out with complete ignorance of the production possibility region. By the process of issuing prices of strategic resources, the CPB acquires incremental knowledge of the feasible set, combining the information received from the sectors, each of which responds with a schedule of feasible output quantities determined in complete independence (or ignorance) of the other sectors' responses. It is possible to say, figuratively, that the western CPB begins the formulation of the plan in a pessimistic state of mind and becomes progressively optimistic as more and more information is exchanged with the sectors.

The entire planning scheme is illustrated in figure 20.1. We assume that the economy is divided into r sectors, each of which may produce some or none of the same commodities of other sectors. In general, every sector will tend to specialize its production, but we wish to indicate that this stringent condition is not necessary for this analysis. The assumption is that the "common good" objective is to maximize, say, the gross national product (GNP). In principle, every sector manager agrees with this objective but may harbor a sectoral perception of that goal and of the contribution of the sector to it.

Resources are classified into strategic and nonstrategic categories. Strategic resources are those essential (basic) inputs required by every sector in the production of its commodities. As an example, oil, coal, electricity, steel, water, and a few other raw materials can be regarded as strategic resources. In general, the list of strategic resources is not very long. Nonstrategic resources are those utilized exclusively by one sector in the production of its outputs. They are specialized (in contrast to basic) resources and their list may be very long, indeed.

At the beginning of the planning process, the CPB possesses the following knowledge: it knows into how many sectors the economy is divided, the total supply of strategic resources, the technology of each sector for employing the strategic resources, and the gross revenue of each sectoral activity. It does not have any knowledge of the supply of specialized resources and neither does it know the technology for their employment. In fact, it does not want to know this information. This is the meaning of decentralized planning, also referred to as planning with limited information at the center.

The planning process is initiated by the CPB, which issues a first round of prices of the strategic resources and waits for the r sectors to respond with their output schedules. Presumably, every sector manager,

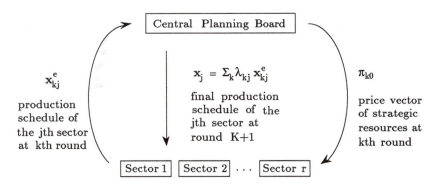

Figure 20.1. The Western scheme of decentralized economic planning.

upon receiving the price information concerning the strategic resources, instructs linear programming experts to compute an optimal production schedule. This schedule is optimal in a sectoral sense. It is not yet known whether it is also optimal or, indeed, even feasible at the economywide level. The information about this production schedule is sent back to the CPB, which adds up all the sectoral demands for strategic resources and compares this total demand with the available total supply of them. In general, the result is that some of the strategic resources are underutilized and some are demanded in excess of the available total supply. This event constitutes an infeasible plan and there is the need for the CPB to issue another round of prices (some higher and some lower than in the previous round) and to wait for a further response from the sector managers.

The exchange of information between the CPB and sectors continues for, say, K rounds until the linear programmers of the CPB determine that an optimal solution for the economy has been reached. At this stage, the CPB issues a last piece of information but, this time, instead of sending a new list of prices for the strategic resources to the sectors, it sends them a schedule of production! As can be seen from figure 20.1, this action of the CPB breaks the symmetry of the process and causes confusion and dismay in some or all of the sector managers. The reason for the managers' surprise can be found in the following question: If, with the last set of prices communicated by the CPB, the sector managers were able to maximize their contribution to the GNP, now that those prices have not been changed, why aren't they allowed to produce the schedule of output quantities that maximizes the sectors' objective functions and was communicated to the CPB in the last correspondence?

The answer to the managers' question is based upon the distinction of the sectors' interest versus the common interest. Because there are limited supplies of strategic resources, the maximum value of the GNP does not correspond with the sum of the optimal values of the sectors' objective functions as determined by the last round of prices. This discrepancy is, in fact, the sole justification for economic planning. Yet, the individual sector managers, who are largely unaware of the other sectors' requests in terms of strategic resources, think that they have done an excellent job in finding the optimal schedule of outputs corresponding to the last set of strategic prices, only to be told that the production plan to be implemented is the one handed down by the CPB. This point is a potential source of conflict between the CPB and at least some of the sectors, and the success or failure of the plan will depend largely on how it is handled.

The logical analysis of the western scheme of economic planning ends here almost without a conclusion. It is not by chance that the detection of possible conflicts is located in correspondence with the symmetry breaking of the information-exchange process. In this setting, symmetry also means harmony, while asymmetry signifies lack of information on the managers' side and the possibility of conflict. At the final $(K + 1)$th stage, sector managers, who initially have accepted the rules of the game (exchange of prices versus production schedules), perceive that those rules have been (arbitrarily?) changed when they are handed both prices and production schedules.

There remains the task of justifying this interpretation of the planning process with a more formal statement of the problem, using a linear programming specification. This is an illustration of linear programming as a way of thinking, in contrast to LP as either a problem-solving procedure or a money-making tool. As stated above, a stylized formulation of an economy's structure will be sufficient to illustrate the essential characteristics of decentralized planning. We begin, therefore, with a specification of the linear programming problem describing the entire economy. This is the problem as we assume it exists prior to the division of tasks between the CPB and the sectors.

The assumptions about the problem are as follows: the economy is divided into r sectors; resources are classified into strategic and specialized inputs; the technology is linear; the objective is to maximize the GNP of the economy. The definition of the model components is:

$A_j \equiv$ the technology matrix for strategic resources of the jth sector
$B_j \equiv$ the technology matrix for specialized resources of the jth sector
$\mathbf{b}_0 \equiv$ the supply vector of strategic resources for the entire economy
$\mathbf{b}_j \equiv$ the supply vector of specialized resources for the jth sector
$\mathbf{x}_j \equiv$ the production schedule of the jth sector

$c_j \equiv$ the unit gross contribution of the jth sector to GNP
$\pi_0 \equiv$ the vector of shadow prices of strategic resources
$\pi_j \equiv$ the vector of shadow prices of the jth sector's specialized resources

The economy program:

Primal

maximize $GNP = c_1'x_1 + c_2'x_2 + \cdots + c_r'x_r$

dual
variables

subject to
$$
\begin{aligned}
A_1x_1 + A_2x_2 + \cdots + A_rx_r &\leq b_0 && \pi_0 \\
B_1x_1 &\leq b_1 && \pi_1 \\
B_2x_2 &\leq b_2 && \pi_2 \\
&\;\;\vdots && \vdots \\
B_rx_r &\leq b_r && \pi_r
\end{aligned}
$$

$$x_j \geq 0, \quad j = 1, \ldots, r$$

We assume that there are m_0 strategic resource constraints and m_j constraints for the specialized resources of the jth sector. As noted above, the number of strategic resources does not need to be very large. In contrast, m_j is usually a very large number. If all the technical and economic information specified by the primal model was known at the level of the CPB, there would be no need to engage in an exchange of information with the sectors and the planning process would be centralized. This type of planning, however, is ruled out because no human effort could collect all the necessary information in a timely fashion.

The economy's primal model exhibits a decomposable structure: if the constraint of strategic resources could be eliminated (or accounted for in an alternative way), the r sector problems would be totally independent of each other. One implausible way to eliminate that constraint and the need for planning is offered by the case where the supply of strategic resources is strictly greater than their total demand. We rule out this possibility, of course. The logical analysis of decentralized economic planning outlined above depends crucially upon a decomposable structure of the problem such as the one presented above (or a similar one). The *decomposition principle* for linear programming, exploited here in the analysis of economic planning, was originally introduced by Dantzig and Wolfe. The dual specification of the economy's problem is

Dual maximize $TC = \mathbf{b}_0' \pi_0 + \sum_{j=1}^{r} \mathbf{b}_j' \pi_j$

$$\text{subject to} \quad A_j' \pi_0 + B_j' \pi_j \geq \mathbf{c}_j$$

$$\pi_0 \text{ and } \pi_j \geq \mathbf{0}, \quad j = 1, \ldots, r$$

Let us consider the viewpoint of a sector manager preparing to formulate the production plan. If the price π_0 of the strategic resources employed by the sector were known, the programming assistants could be instructed to solve the following LP sectoral problem.

The jth sector problem:

Primal maximize $VA = (\mathbf{c}_j' - \pi_0' A_j)\mathbf{x}_j$

$$\text{subject to} \quad B_j \mathbf{x}_j \leq \mathbf{b}_j$$

$$\mathbf{x}_j \geq \mathbf{0}$$

The dual objective function is to be interpreted as the value added (VA) of the jth sector, since it is the difference between gross receipts and the cost of strategic raw materials. When the prices of the strategic resources (π_0) become known, the objective function is linear and the problem is ready for solution. The dual sectoral problem is

Dual maximize $CSR = \mathbf{b}_j' \pi_j$

$$\text{subject to} \quad B_j' \pi_j \geq \mathbf{c}_j - A_j' \pi_0$$

$$\pi_j \geq \mathbf{0}$$

The dual objective function minimizes the cost of the specialized resources (CSR), while the dual constraints state that the marginal cost of the specialized resources must be greater than or equal to the marginal value added for the sector. We assume that the feasible region of the jth sectoral problem is bounded and nonempty. Suppose it looks as shown in figure 20.2. The feasible region is a convex polygon (set) with K_j extreme points. Then, it is a property of a convex polygon that any point in its interior can be expressed as a convex combination of some, or all, of its extreme points. Hence, we can infer that any feasible solution of the jth sectoral primal problem, $\mathbf{x}_j$, can be expressed as a convex combination of all the K_j extreme points, $\mathbf{x}_{kj}^e$, as illustrated in figure 20.2. In a formal notation, the convex combination of the jth feasible solution is expressed

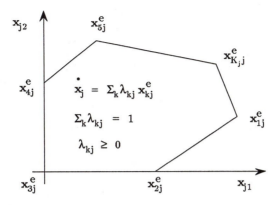

Figure 20.2. The feasible region of the jth sector.

as

$$\mathbf{x}_j = \sum_{k=1}^{K_j} \lambda_{kj} \mathbf{x}_{kj}^e$$

with

$$\sum_{k=1}^{K_j} \lambda_{kj} = 1 \quad \text{and} \quad \lambda_{kj} \geq 0, \qquad k = 1, \ldots, K_j$$

The coefficients λ_{kj} are simply nonnegative weights used to aggregate the basic feasible solutions, $\mathbf{x}_{kj}^e$, which, we know, correspond to extreme points. Also the linear programmers working for the CPB know this much and will use their knowledge accordingly to define a linear programming problem of the CPB, known as the *master program*.

The division of information between the sectors and the CPB assigns to the CPB the constraints involving the strategic resources. Furthermore, when the production schedules, $\mathbf{x}_{kj}^e$, from the various sectors arrive at the center, they can be used to define the convex combination of extreme points as discussed. The linear problem of the CPB can, therefore, be stated as given below.

The master program:

Primal   maximize $GNP = \sum_{j=1}^{r} \sum_{k=1}^{K_j} \lambda_{kj} [c'_j x^e_{kj}]$

dual
variables

subject to $\sum_{j=1}^{r} \sum_{k=1}^{K_j} \lambda_{kj} [A_j x^e_{kj}] \leq b_0$ π_0

$\sum_{k=1}^{K_j} \lambda_{kj} = 1$ σ_j

$\lambda_{kj} \geq 0$

$j = 1, \ldots, r,$ and $k = 1, \ldots, K_j$

The master program is obtained by replacing the x_j quantities in the economy model by their equivalent expressions in terms of the convex combinations of the extreme points of each sector's problem. The constraints involving the specialized resources no longer concern the CPB, since they are of exclusive pertinence for each sector's manager. In their place, the master program exhibits one constraint for every sector, which insures that the weights λ_{kj} add up to 1. This constraint is associated with the definition of the linear combination discussed above. The quantities in the brackets are known to the CPB as soon as every sector communicates its production schedule x^e_{kj}. The unknowns of the master program, therefore, are the weights λ_{kj} to be used in the computation of the weighted average (convex combination) of the production schedules of each sector. The number of constraints of the master program is much smaller than the total number of constraints for the economy. There are m_0 constraints involving the strategic resources and only r constraints expressing the sum to unity of the weights λ_{kj}. The primal version of the master program, therefore, seeks to find the optimal weights for all the production schedules received by the CPB during the process of formulating the economic plan.

The dual master program is

Dual   minimize $TC = b'_0 \pi_0 + \sum_{j=1}^{r} \sigma_j$

subject to $[A_j x^e_{kj}] \pi_0 + \sigma_j \geq c'_j x^e_{kj}$

$\pi_0 \geq 0,$ σ_j free, $j = 1, \ldots, r,$ and $k = 1, \ldots, K_j$

The interpretation of the dual master program is the minimization of the total imputed value of the strategic resources ($b'_0 \pi_0$) plus the imputed value added of all the r sectors. This interpretation of the sum of all σ_j's is supported by the structure of the dual constraints, which are conveniently rewritten as

$$\sigma_j \geq (\mathbf{c}_j' - \pi_0' A_j)\mathbf{x}_{kj}^e$$

indicating that σ_j has a lower bound formed by the value added of the jth sector. The dual variable σ_j is free because it is associated with a primal constraint in the form of an equation. The economic consequence of this unrestricted variable is that it is possible for an optimal plan to include a sector exhibiting a negative value added. Although this conclusion may seem counterintuitive, it is nonetheless admitted by the assumptions specified initially: another instance of how common sense is labile and, in science, rather worthless.

The planning process can now begin with the CPB issuing a set of prices π_0 for the strategic resources. In the initial round, for example, all the prices could be zero. Since, at the beginning, the western CPB is almost totally ignorant, a set of zero prices can be interpreted as a wait-and-see attitude: the sectors have maximum freedom of choice in the employment and demand of strategic resources. This is an easy way to gauge an upper bound on their total demand.

The sector managers, upon receiving π_0, solve the sectoral LP problem and send the optimal solution to the CPB. This optimal solution must correspond to an extreme point of the feasible region, the only reason for using the notation $\mathbf{x}_{kj}^e$. With this information, the CPB calculates the quantities $\mathbf{c}_j'\mathbf{x}_{kj}^e$ and $A_j\mathbf{x}_{kj}^e$ for each sector and uses them as known coefficients in the master program whose solution produces a set of weights, λ_{kj}, a new vector of shadow prices for the strategic resources, π_0, and a set of imputed value added parameters, σ_j. The weights and the values added are kept at the CPB and are not communicated to the sector managers, since they are temporary parameters and, at the intermediate stages of the planning process, are not essential for the exchange of the crucial information between the center and the periphery. Only the new set of prices for the strategic resources reach the sectors, and the process iterates once again. As more and more production schedules are communicated by the sectors to the CPB during the various rounds, the agency becomes more and more knowledgeable about the feasible region of the economy; the initial pessimism is replaced by an increasing optimism.

The stopping rule is in the hands of the CPB and corresponds to the familiar optimality criterion for a linear program according to the primal simplex algorithm: when all the opportunity costs are nonnegative, an optimal solution has been found.

Let us recall that, from section 7.3, the dual definition of opportunity cost for a general linear programming problem is $(z_j - c_j) = \mathbf{a}_j'\mathbf{y} - c_j$. The master program exhibits the use of two indexes, j and k, and, therefore, the appropriate notation for the opportunity costs of this problem is

$$
\begin{aligned}
z_{kj} - f_{kj} &= [A_j \mathbf{x}_{kj}^e]' \pi_0 + \sigma_j - \mathbf{c}_j' \mathbf{x}_{kj}^e \\
&= (\pi_0' A_j - \mathbf{c}_j') \mathbf{x}_{kj}^e + \sigma_j \\
&= -VA_{kj} + \sigma_j
\end{aligned}
$$

where $z_{kj} \equiv [A_j \mathbf{x}_{kj}^e]' \pi_0 + \sigma_j$, and $f_{kj} \equiv \mathbf{c}_j' \mathbf{x}_{kj}^e$. To facilitate the assimilation of this expression, we notice that the right-hand side is nothing else than the constraint of the dual master program. When the $(z_{kj} - f_{kj})$ will be nonnegative for all k and j, an optimal economic plan will be found. At that stage, the optimal weights λ_{kj} will be used to define the optimal (for the economy and not necessarily for the sectors) convex combinations of the production schedules communicated to the CPB. The quantity vectors $\mathbf{x}_j$ represent the commodity levels that the sectors should produce in order to maximize the GNP, but they may represent interior points of some or all sectors' feasible regions. In other words, from the viewpoint of a sector manager, the CPB's $\mathbf{x}_j$ is not an optimal schedule because it does not correspond to an extreme point of the perceived feasible region. This is the moment when consensus may break down together with the symmetry of the process.

20.4 The Russian Scheme of Economic Planning

For several decades, the Soviet Union adopted an economic planning process called "planning in physical terms." During those same years, prices played a minor role, if any, in the decisions of the Russian Central Planning Board. The process has apparently changed in recent years, but the folklore remains. In any case, we do not wish to discuss the actual planning process adopted in the Soviet Union. We borrow only the idea of a CPB committed to formulating an optimal plan for the economy by prodding the various sectors with a series of requests for the production of certain levels of outputs using certain levels of resources. For this reason, the process has already been identified as the dual of the western scheme of planning discussed in the previous section. As a dual specification, it represents attitudes and procedures that are opposite, in some sense, to those encountered in the western scheme.

Consider the illustration of the scheme of economic planning given in figure 20.3. The planners of the Russian central board begin the process with an attitude of great optimism about the feasible region of the economy. In fact, it is the belief of knowing it all that allows them to issue quotas $(\mathbf{u}_{kj})$ of products and resources to the various sectors of the economy. Some

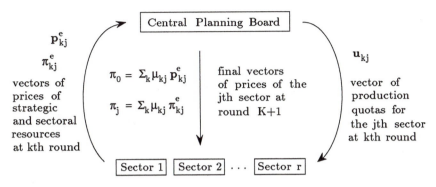

Figure 20.3. The Russian scheme of decentralized economic planning.

(or all) of those quotas soon are shown to be infeasible by the sectors' managers, who possess the detailed knowledge of the technology required to match resources and products. Their only possibility of communicating with the CPB is via the calculation of the shadow prices of those quotas, $\mathbf{p}^e_{kj}$ and π^e_{kj}. Now it happens that the same resource used by different sectors gives rise to different shadow prices. This is a sign of inefficient allocation of resources and, for this reason, the CPB must modify the quota assigned to each sector. Furthermore, prices define hyperplanes, and we can interpret the shadow prices communicated by the sectors as defining cutting planes for the utopian notion of a feasible region entertained initially by the CPB. Through the iterated execution of this exchange process, the Russian CPB is brought down to reality, as far as technical possibilities are concerned, by a series of eliminations of the initially dreamed feasible region.

When the same resource used in different sectors obtains identical shadow prices, an optimal plan has been reached. At this stage the CPB has a final communication with the sector managers, who are notified of a set of uniform prices rather than quotas. This action corresponds to the asymmetry already encountered in the western scheme that causes the same possibility of misunderstanding. Thus, in the end, both schemes of decentralized economic planning face the same difficult task of convincing the personnel in charge of implementing the plan that the final instructions that seem so strange are, in fact, an essential part of the process. If political and psychological aspects such as ideology, greed, bad faith, fear, and cheating are superimposed on this logical conclusion, it is not difficult to understand the improbable odds facing an economic plan.

The formal specification of the Russian scheme of decentralized economic planning begins with a restatement of the economy problem. In this case, however, new variables $\mathbf{u}_j$, called quotas, are introduced. These quantity vectors represent the strategic commodity levels assigned to each sector.

The economy problem:

Primal maximize $GNP = \sum_{j=1}^{r} \mathbf{c}_j' \mathbf{x}_j$

dual
variables

subject to

$$\sum_{j=1}^{r} \mathbf{u}_j \leq \mathbf{b}_0 \qquad \pi_0$$

$$A_j \mathbf{x}_j - \mathbf{u}_j \leq \mathbf{0} \qquad \mathbf{p}_j$$

$$B_j \mathbf{x}_j \leq \mathbf{b}_j \qquad \pi_j$$

$$\mathbf{x}_j \geq \mathbf{0} \text{ and } \mathbf{u}_j \text{ free}, \qquad j = 1, \ldots, r$$

The objective is always to maximize the GNP. The first constraint specifies that the sum of the quotas must add up to either the available supply of strategic resources or to the required demand of strategic final products. The second constraint states that the demand of the jth sector for strategic commodities, $A_j \mathbf{x}_j$, must be less than or equal to the supply of them, $\mathbf{u}_j$. The third set of constraints represents the technology of specialized resources. The quotas are free variables to allow for the possibility of disinvestment. This assumption is also required for guaranteeing the equality of shadow prices across sectors, as indicated in the dual program that now follows:

Dual minimize $TC = \mathbf{b}_0' \pi_0 + \sum_{j=1}^{r} \mathbf{b}_j' \pi_j$

primal
variables

subject to

$$A_j' \mathbf{p}_j + B_j' \pi_j \geq \mathbf{c}_j \qquad \mathbf{x}_j$$

$$\pi_0 - \mathbf{p}_j = \mathbf{0} \qquad \mathbf{u}_j$$

$$\pi_0, \pi_j \text{ and } \mathbf{p}_j \geq \mathbf{0}, \qquad j = 1, \ldots, r$$

The dual objective function is the familiar minimization of the imputed cost of the strategic commodities and the specialized resources. The first set of constraints specifies that, for each sector, the marginal cost of the activities must be greater than or equal to the marginal gross revenue (GR). The second set of constraints guarantees the equality of the shadow prices of the strategic commodities across all sectors. Finally, all the shadow prices must be nonnegative.

This economywide specification must now be partitioned between the CPB and the sectors. We begin with the jth sector. As soon as the quota $\mathbf{u}_j$ is announced, the jth sector problem assumes the following format.

The jth sector problem:

Primal maximize $TC_j = \mathbf{u}'_j \mathbf{p}_j + \mathbf{b}'_j \pi_j$

$$\begin{array}{c} \text{dual} \\ \text{variables} \end{array}$$

subject to $\qquad A'_j \mathbf{p}_j + B'_j \pi_j \geq \mathbf{c}_j \qquad \mathbf{x}_j$

$$\mathbf{p}_j \text{ and } \pi_j \geq \mathbf{0}$$

The minimization of the total cost is selected as the primal sectoral problem because the sectors must respond to the CPB with shadow prices. The dual specification is

Dual maximize $GR_j = \mathbf{c}'_j \mathbf{x}_j$

subject to $\qquad A_j \mathbf{x}_j \leq \mathbf{u}_j$

$$B_j \mathbf{x}_j \leq \mathbf{b}_j$$

$$\mathbf{x}_j \geq \mathbf{0}$$

As for the jth sectoral problem of the western scheme, we assume that it is possible to express an interior point of the primal feasible region as a convex combination of all (or some) H_j extreme points, that is,

$$\mathbf{p}_j = \sum_{h=1}^{H_j} \mu_{hj} \mathbf{p}^e_{hj}$$

$$\pi_j = \sum_{h=1}^{H_j} \mu_{hj} \pi^e_{hj}$$

with

$$\sum_{h=1}^{H_j} \mu_{hj} = 1 \quad \text{and} \quad \mu_{hj} \geq 0, \qquad h = 1, \dots, H_j$$

This assumption (guaranteed, in general, rather easily) is important for the programmers of the CPB who can now state their master programs as follows.

The master program:

Primal minimize $TC = \mathbf{b}_0'\pi_0 + \displaystyle\sum_{j=1}^{r}\sum_{h=1}^{H_j}\mu_{hj}[\mathbf{b}_j'\pi_{hj}^e]$

dual
variables

subject to $\quad \pi_0 - \displaystyle\sum_{h=1}^{H_j}\mu_{hj}\mathbf{p}_{hj}^e = 0 \qquad \mathbf{u}_j$

$\displaystyle\sum_{h=1}^{H_j}\mu_{hj} = 1 \qquad \phi_j$

$\mu_{hj} \geq 0$

$h = 1, \ldots, H_j$, and $j = 1, \ldots, r$. The primal problem of the CPB is, again, to find optimal weights, μ_{hj}, for eventually defining the optimal shadow prices of the strategic commodities. When the information pertaining to the shadow prices of the strategic and specialized resources arrives at the CPB headquarters from every sector, it is treated as a known parameter for the master program, which is then solved for the optimal weights. The associated dual master program furnishes the revision of the quotas:

Dual maximize $GNP = \displaystyle\sum_{j=1}^{r}\phi_j$

subject to $\qquad \displaystyle\sum_{j=1}^{r}\mathbf{u}_j = \mathbf{b}_0$

$\phi_j - \mathbf{u}_j'\mathbf{p}_{hj}^e \leq \mathbf{b}_j'\pi_{hj}^e$

ϕ_j and $\mathbf{u}_j$ free

$h = 1, \ldots, H_j$, and $z_j = 1, \ldots, r$. As for the western scheme, the stopping rule for the master program is furnished by the dual constraints, which require the feasibility of the quotas $\mathbf{u}_j$ and $\phi_j \leq TC_{hj}$. In economic words, the imputed total cost for each sector must be greater than or equal to the imputed contribution of that sector to the GNP. At that stage, the CPB of the Russian scheme of decentralized planning should be ready to face the sector managers and to convince them that it is in the common interest to accept both quotas and prices as determined at the CPB headquarters.

20.5   A Comparison of Planning Schemes

While the master program of the western scheme is centered around a maximization objective, that of the Russian scheme specifies a minimization goal. The same duality applies to sectors. It is interesting to notice that we have been talking about dual schemes of economic planning (the western and the Russian), each of which is, in turn, specified in terms of dual master programs. This nested duality structure suggests the possibility of establishing a horizontal relationship between the components of one scheme and those of the other one. At the optimum, for example, the primal objective function of the western scheme should be equal to the dual objective function of the Russian scheme. This allows us to give an explicit economic meaning to the sum of the ϕ_j variables, which now can be related to the western GNP:

$$\sum_{j=1}^{r} \phi_j = \sum_{j=1}^{r} \sum_{k=1}^{K_j} \lambda_{kj} [\mathbf{c}_j' \mathbf{x}_{kj}^e]$$

Furthermore, it must be that

$$\sum_{j=1}^{r} \sigma_j = \sum_{j=1}^{r} \sum_{h=1}^{H_j} \mu_{hj} [\mathbf{b}_j' \pi_{hj}^e]$$

Which of the two schemes of economic planning imposes the minimal information requirements upon the respective CPBs? The western scheme requires knowledge of the following elements: $\mathbf{b}_0$, $\mathbf{c}_j$, and A_j, that is, the available quantities of strategic resources, the gross revenue of each activity for every sector, and the matrix of technology that employs strategic resources. The Russian scheme requires knowledge of $\mathbf{b}_0$ and $\mathbf{b}_j$, the total of all commodity quotas and the supply of specialized resources for each sector. At first sight it would seem that the Russian scheme is to be preferred, but one should not discount the difficulty of collecting the supply information about the specialized resources. It may be easier to know the technological matrix A_j, since it is usually defined by engineers. The final answer may have to be judgmental.

It is impossible to conclude this chapter and this book without a final, admiring glance at the symmetry revealed by the two schemes of decentralized economic planning. Since not even the opponents deny the relevance of the subject, there is hope that by focusing on the symmetric structure of either scheme in isolation and on the larger symmetry resulting from their combination, further insights can be uncovered to give economic planning a possibility of success. Among all the intriguing symmetries, however, we cannot forget the crucial asymmetry that weakens the predictive power of either scheme. It is upon this broken symmetry that further attention must be focused in order to find a more general specification of the planning problem within which this last asymmetry can also be resolved.

On a purely logical ground, a strategy for overcoming the asymmetric aspect of decentralized planning and the associated conflicts suggests that the planning process be organized both as a Western and a Russian scheme. Although this approach would require that the CPB possess additional information at the beginning of the exchange, it would give the sectors' managers a dual reference scheme to gauge their contribution to the common goal. In this way, the planning process itself provides the framework for understanding the origin of possible conflicts and their resolution.

Bibliography

Boles, James N., "The 1130 Farrell Efficiency System–Multiple Products and Multiple Factors," Giannini Foundation of Agricultural Economics, University of California, Berkeley, 1971.

Boole, George, *A Treatise On Differential Equations*, New York, Chelsea Pub. Co., fifth edition, 1859.

Dantzig, George B., "Programming in a Linear Structure," Comptroller, USAF, Washington, D.C., February 1948.

Dantzig, George B., and Philip Wolfe, "Decomposition Principle for Linear Programs," *Operations Research*, vol. 8, January-February 1960, pp. 101–111.

Farrell, Michael J., "The Measurement of Productive Efficiency," *Journal of the Royal Statistical Society*, Series A, Part III, vol. 120, 1957, pp. 253–281.

Friedman, Julia, "A Comparative Cost Analysis of the Production of Energy from Solid Waste in Oregon," Human Resources and Natural Resources Committees of the Western Agricultural Economics Research Council, Economic Impacts of Rapid Development of Energy Resources, Methodology and Issues, Report n. 5, 1974, pp. 15–31.

Goldman, A. J., and A. W. Tucker, "The Theory of Linear Programming," in *Linear Inequalities and Related Systems*, edited by H. W. Kuhn and A. W. Tucker, Princeton, Princeton University Press, 1956, pp. 53–98.

Hitchcock, Frank L., "The Distribution of a Product from Several Sources to Numerous Localities," *Journal of Mathematics and Physics*, vol. 20, 1941, pp. 224–230.

Kantorovich, L. V., "Mathematical Methods in the Organization and Planning of Production," Publication House of the Leningrad State University, Leningrad, 1939, 68 pp. Translated in *Management Science*, vol. 6, 1960, pp. 366–422.

Karmarkar, Narendra, "A New Polynomial-Time Algorithm for Linear Programming," *Combinatorica*, vol. 4, 1984, pp. 373–395.

Koopmans, T. C., "Optimum Utilization of the Transportation System," *Proceedings of the International Statistical Conference*, 1947, Washington, D.C., vol. 5. Reprinted as Supplement to *Econometrica*, vol. 17, 1949.

Lemke, C. E., "The Dual Method for Solving the Linear Programming Problem," *Naval Research Logistic Quarterly*, vol. 1, 1954, pp. 36–47.

Rybczynski, T. M., "Factor Endowment and Relative Commodity Prices," *Econometrica,* vol. 22, 1955, pp. 336–341.

Stigler, George J., "The Cost of Subsistence," *Journal of Farm Economics,* vol. 27, May 1945, pp. 303–314.

Stolper, W. F., and P. A. Samuelson, "Protection and Real Wages," *Review of Economic Studies,* vol. 9, 1941, pp. 58–73.

Weitkamp, W. H., Clawson, W. J., Center, D. M., and W. A. Williams, "A Linear Programming Model for Cattle Ranch Management," Division of Agricultural Sciences, University of California, Bulletin 1900, December 1980.

Glossary

Activity: a series of coefficients arranged in a column and describing the technical requirements for transforming commodity inputs into commodity outputs. A synonymous for activity is process.

Algorithm: a complete procedure (method) describing the steps to be taken for the solution of a problem. Four algorithms are discussed in this book: primal simplex, dual simplex, artificial variable and artificial constraint algorithm.

Basis: a set of linearly independent vectors spanning the given space. Alternatively, a reference system for locating (measuring) objects in a given space. Geometrically it is represented by a cone in the primal input requirement space.

   primal basis: a basis for the primal LP problem.

   dual basis: a basis for the dual LP problem.

   feasible basis: a basis that generates a basic feasible solution.

   infeasible basis: a basis that generates an infeasible solution.

Ceteris paribus: a logical procedure for studying the behavior of a system as only one element of the problem is allowed to vary, with all the other elements kept at their initial values.

Commodity: any good, whether input or output, of economic interest.

Complementary slackness condition: the nullity condition associated with the product of an inequality constraint and its corresponding dual variable (Lagrange multiplier). There are primal as well as dual complementary slackness conditions. They form the necessary and sufficient conditions for the optimality of LP problems.

Cone: a geometrical figure generated by vectors emanating from the origin of a Cartesian coordinate system with less than 180 degree angle among them.

Constraint: a linear relation defining the structure of a LP problem. It can be either an equation or an inequality.

Convex combination: the weighted sum of vectors performed with nonnegative weights that sum to one.

Criterion: a numerical test performed during a solution procedure.

   entry criterion: a test for selecting the activity to enter the basis in the next iteration of the simplex algorithms.

   exit criterion: a test for identifying the activity that must leave the basis in the next iteration of the simplex algorithms.

feasibility criterion: a test performed during the simplex algorithms by means of a minimum ratio computation for guaranteeing the feasibility of the next basic feasible solution.

optimality criterion: a test performed on the opportunity cost coefficients to determine whether the current basic feasible solution is optimal.

Decomposition principle: a mathematical notion used in decentralized planning according to which the problem of formulating a plan can be decomposed into subproblems associated with the center and the periphery (sectors).

Degeneracy: a mathematical property of a basic solution according to which one or more basic variables take on a zero value. In a two-dimensional space, for example, degeneracy is represented geometrically by a point that is the intersection of three or more lines. In LP we may have *primal* as well as *dual degeneracy*. Degeneracy of primal and/or dual solutions has a direct link with the existence of dual and/or primal *multiple optimal solutions*.

Dual: an attribute of *space, problem, constraint, region, variable, basis, solution*.

Duality: the property of mathematical problems relating prices and quantities, variables and parameters, constraints and Lagrange multipliers.

Equilibrium conditions: the necessary relations for the proper specification of primal and dual problems. For example, in a LP specification of the competitive firm, the primal equilibrium conditions are given by: *demand of commodities* $\leq$ *supply of commodities*. The dual equilibrium conditions for any activity are: *marginal cost* $\geq$ *marginal revenue*.

Extreme point: a corner point in a convex polyhedral figure; that is, a point on the boundary of a convex set that cannot be expressed as a convex combination of any set of distinct points belonging to the set.

Foregone income: the income loss due to the release of resources necessary for investing in an alternative productive activity. Income foregone has two definitions: a primal definition at market prices and a dual definition at factor costs.

Free good: a commodity in excess supply (surplus) relative to its demand. A free good is associated with a zero price.

Lagrangean function: the relation defined in the process of solving an optimization problem subject to constraints.

Lagrange multiplier: the variable associated with a constraint in the Lagrangean function. A Lagrange multiplier is, thus, a dual variable.

Linear independence (dependence): a property of vectors in a given space. Geometrically, non-parallel vectors are linearly independent; parallel vectors are linear dependent.

Linear programming: a mathematical structure of optimization problems whose objective function and constraints are linear relations.

Linear relation: any function, equality and/or inequality constraint defined by variables entering in a linear way as opposed to products, logarithms, powers, etc.

Marginal rate of technical transformation: the ratio of derivatives of a function with respect to two variables, with all the other variables kept at given levels. In technical terms it measures by how much a commodity should be decreased (increased) in order to obtain a unit increase (decrease) of a second commodity.

Marginal rate of substitution: the marginal rate of technical transformation between two commodity inputs.

Marginal product: the marginal rate of technical transformation between a commodity input and a commodity output.

Master program: the subproblem associated with the central planning board in a decentralized planning process.

Matrix: an array of numbers arranged in rows and columns.

    transformation matrix: a square matrix defined at each iteration in the pivot method for solving systems of linear equations.

    identity matrix: a square matrix with unit values on the main diagonal and zero everywherelse. Geometrically, it corresponds to the origin of a space in a Cartesian coordinate system.

    inverse matrix: a square matrix generated (either implicitly or explicitly) during the solution of a basic system of equations, with the property that the product of such an inverse matrix and the matrix of coefficients of the basic system is equal to the *identity matrix.*

Minimum ratio: the criterion used in the simplex algorithms for maintaining the feasibility of a basic solution.

Objective function: in a LP problem, a linear relation to be either maximized or minimized.

Opportunity cost: a major economic criterion for judging the profitability of an activity. Defined as the difference between *foregone income* and the marginal revenue of the activity to be evaluated.

Opportunity input requirement: a criterion for judging the technical efficiency of an activity. Analogous to the opportunity cost but evaluated only in physical terms.

Orthogonal (normal, perpendicular) vectors: two vectors are orthogonal if and only if they form a 90 degree angle between them (if and only if their inner product is equal to zero).

Parametric programming: a post-optimality analysis for studying the behavior of the optimal solution as some parameter of the given problem is varied over an admissible domain. The variation of the right-hand-side vector of constraints allows the derivation of input demand functions. The variation of the objective function's vector of coefficients allows the derivation of output demand functions. See also *sensitivity analysis.*

Phase I: the stage of a solution procedure devoted to finding (if it exists) a *basic feasible solution.* It is associated with the *artificial variable algorithm.*

Phase II: the stage of a solution procedure devoted to finding an **optimal solution** (if it exists). It is associated with either the *primal* or the *dual simplex algorithm.*

Pivot algorithm (method): a solution procedure for solving a system of linear equations.

   pivot element: the element selected at each iteration of the pivot algorithm as a seed for defining the corresponding *transformation matrix.*

Planning: the procedure of formulating a strategy; usually a production strategy. Formulating a strategy requires the availability of information. *Centralized planning* is the process of formulating a strategy when all the information resides at the center. *Decentralized planning* is the procedure of formulating a strategy when the center has available only limited information and must interact with the periphery *(sectors)* in order to collect the remaining information.

   central planning board: the central agency charged with formulating a plan. Sectors, or divisions, are the peripheral units interacting with the central planning board in the exchange of information for the formulation of the plan.

   planning scheme: for convenience, the approaches to planning are classified in two groups: the *Western scheme* and the *Russian scheme.* The Western scheme is based upon the use of prices by the central planning board in the transmission of information to the sectors; the Russian scheme is based upon the use of quotas by the central planning board in its information exchanges with the sectors.

   quotas: specified assignments of physical amounts of inputs and outputs communicated by the central planning board to the sectors.

Primal: an attribute associated with several notions such as *space, problem, constraint, region, variable, basis, solution.*

Process: in this book, a synonymous for *activity.*

Sectoral program: the subproblem associated with an individual sector in a decentralized planning process.

Sensitivity analysis: the study of the stability of the *optimal basis* in relation to the variation of some parameter of the given LP problem.

Simplex: a geometrical figure defined by $(n+1)$ points in a n-dimensional space with the condition that not all $(n+1)$ points lie on the same hyperplane. For example, in a 2-dimensional space, 3 points not arranged on the same line form a triangle: a simplex. In a 3-dimensional space, a tetrahedron (a four-edge object) forms a simplex.

simplex algorithm: a very successful method for solving LP problems invented primarily by G. B. Dantzig. At each iteration the method constructs a simplex in the appropriate space.

Solution: a set of values assumed by the variables in a system of equations (inequalities) that satisfy the given equations (inequalities).

basic solution: a solution to a system of equations whose elements associated with basic vectors may be different from zero while the elements associated with non-basic vectors are equal to zero.

basic feasible solution: for primal LP problems, a basic solution whose elements associated with basic vectors are nonnegative. In dual LP problems, a basic feasible solution may have unrestricted basic variables.

nonbasic feasible solution: for primal LP problems, a solution whose elements are all nonnegative. The elements associated with non-basic vectors are not restricted to be equal to zero.

optimal basic feasible solution: a basic feasible solution associated with the optimal value of the objective function.

near- (quasi-) optimal solution: either basic feasible solutions or non-basic feasible solutions associated with a value of the objective function very close to the optimal value.

multiple optimal solutions: distinct solutions, either basic feasible or non-basic feasible, associated with the same optimal value of the objective function.

degenerate solution: a basic feasible solution with one (or more) element associated with a basic vector that is equal to zero.

unbounded solution: a basic feasible solution associated with an unbounded value of the objective function.

infeasible solution: a set of values for the variables of the given problem that violates one (or more) constraint(s).

Space: the universe of admissible representations of a LP problem.

primal spaces: spaces of representation of primal LP problems (output space and input requirement space).

dual spaces: spaces of representation of dual LP problems (spaces of output and input prices).

output space: primal space spanned by primal variables (output quantities in a firm interpretation).

input requirement space: primal space spanned by primal constraints (input quantities in a firm interpretation).

Symmetry: the property of mathematical objects (problems) describing their similar (invariant) structure.

System of linear relations: a set of linear equations and inequalities characterizing the structure of LP constraints.

basic system: a set of linear equations generating a basic solution.

nonbasic system: a set of linear equations generating a non-basic solution.

decomposable system: a set of linear relations admitting the possibility of solution by means of subsystems; used in decentralized planning.

Technology: the set of relations describing the transformation of commodities by means of commodities. Referred also as a production function. Conventionally, a technology is described by a matrix whose column vectors are interpreted as activities.

Variable: an element of a LP problem whose value is indeterminate prior to the problem's solution.

primal variable: a variable of a primal LP problem, denoted by the symbol x in this book.

dual variable: a variable of dual LP problem, denoted by the symbol y in this book.

slack variable: a nonnegative variable associated with either primal or dual inequality constraints. In economic terms it is called a surplus variable.

artificial variable: a nonnegative variable introduced in a LP problem for the purpose of finding a basic feasible solution (if it exists).

nonnegative variable: a variable required to assume only nonnegative values.

free variable: an unrestricted variable, that is a variable that can assume any value, from negative to positive infinity.

Vector: a geometrical representation of a point in a linear space using the origin of a Cartesian coordinate system as a reference point. An activity (or process) is also regarded as a vector in the corresponding space.

Index

333